Plant Processing from a Prehistoric and Ethnographic Perspective

Préhistoire et ethnographie du travail des plantes

Plant Processing from a Prehistoric and Ethnographic Perspective

Préhistoire et ethnographie du travail des plantes

Proceedings of a workshop at Ghent University (Belgium) November 28, 2006

Edited by

Valérie Beugnier and Philippe Crombé

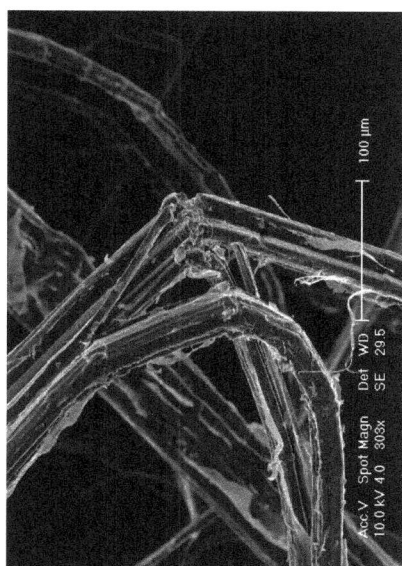

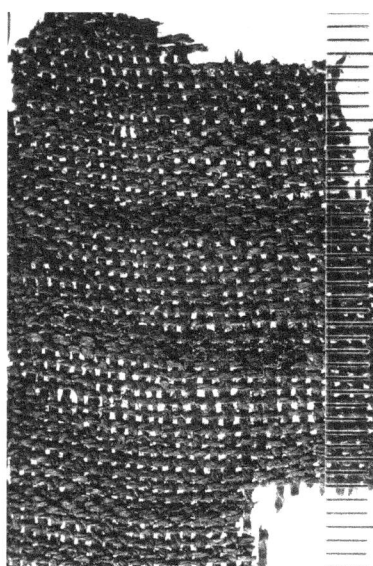

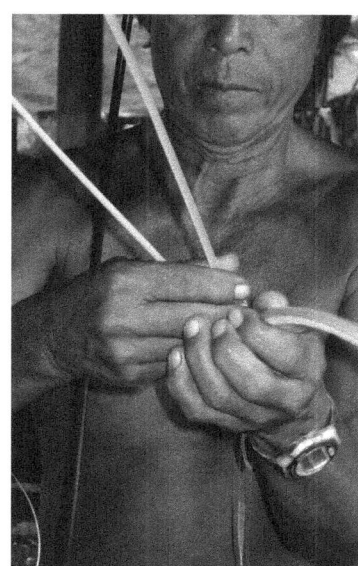

BAR International Series 1718
2007

Published in 2016 by
BAR Publishing, Oxford

BAR International Series 1718

Plant Processing from a Prehistoric and Ethnographic Perspective
Préhistoire et ethnographie du travail des plantes

ISBN 978 1 4073 0201 0

BAR Publishing is the trading name of British Archaeological Reports (Oxford) Ltd.
British Archaeological Reports was first incorporated in 1974 to publish the BAR
Series, International and British. In 1992 Hadrian Books Ltd became part of the BAR
group. This volume was originally published by John and Erica Hedges Ltd. in
conjunction with British Archaeological Reports (Oxford) Ltd / Hadrian Books Ltd, the
Series principal publisher, in 2007. This present volume is published by BAR
Publishing, 2016.

Printed in England

BAR
PUBLISHING

BAR titles are available from:

BAR Publishing
122 Banbury Rd, Oxford, OX2 7BP, UK
EMAIL   info@barpublishing.com
PHONE   +44 (0)1865 310431
FAX   +44 (0)1865 316916
www.barpublishing.com

CONTENTS

Acknowledgements/Remerciements   *3*

Introduction.
Plant processing from a prehistoric and ethnographic perspective/Préhistoire et
ethnographie du travail des plantes.
Valérie Beugnier and Philippe Crombé   *5*

Where would we be without string?
Ethnographic and prehistoric evidence for the use, manufacture and role of string
in the Upper Palaeolithic and Mesolithic of Northern Europe.
Karen Hardy   *9*

Préhistoire du travail des plantes dans le nord de la Belgique.
Le cas du Mésolithique ancien et du Néolithique final en Flandre.
Valérie Beugnier   *23*

Plant processing for cordage and textiles using serrated flint edges:
new *chaînes opératoires* suggested by combining ethnographic,
archaeological and experimental evidence for bast fibre processing.
Linda Hurcombe   *41*

Acquisition et traitement des matières textiles d'origine végétale en Préhistoire :
l'exemple du lin.
Emmanuelle Martial et Fabienne Médard   *67*

Investigating social aspects of technical processes: cloth production
from plant fibres in a Neolithic lake dwelling on Lake Constance, Germany.
Susanna Harris   *83*

Aroumans (*Ischnosiphon* spp., Marantaceae).
Vannerie et symbolisme en Guyane française.
Damien Davy   *101*

Acknowledgements – Remerciements

First, we would like to thank Ghent University who ensured, in November 2006, that the round table took place in the best conditions. We also thank all the participants of the meeting, some of whom were unable to take part in this publication. We sincerely thank the authors who contributed to this volume for the quality of their work, their enthusiasm and their encouragement. Finally, we would like to express our warmest thank to Micheline De Wit (ASBL ADIA, Institut royal des Sciences naturelles de Belgique) who contributed to the editing process with know-how, patience and kindness.

***

Tout d'abord, nous souhaitons remercier l'Université de Gand qui a permis l'organisation dans des conditions idéales de la table ronde dont les actes sont publiés ici. Un grand merci également à l'ensemble des participants de ces rencontres, dont certains n'ont pu se joindre à la publication. Nous adressons également nos plus vifs remerciements aux auteurs dont l'enthousiasme, les encouragements et la qualité des contributions sont à l'origine de ce projet. Toute notre gratitude va également à Micheline De Wit (ASBL ADIA, Institut royal des Sciences naturelles de Belgique) qui a assuré avec patience, adresse et gentillesse la mise en forme de cet ouvrage.

Valérie Beugnier, Philippe Crombé

INTRODUCTION
PLANT PROCESSING FROM A PREHISTORIC
AND ETHNOGRAPHIC PERSPECTIVE

Valérie BEUGNIER and Philippe CROMBÉ

Recently, different researches have demonstrated that plants play an important part in the life of prehistoric and traditional societies (Heider 1970, Pétrequin et Pétrequin 1988, Sillitoe 1988, Coles and Coles 1989, Owen 1993, Barber 1994, Adovasio *et al.* 2007). In these groups, plants constitute a major component not only in the subsistence but also in the manufacture of all kinds of tools, weapons, objects, traditional medicine and art or ritual products. A great variety of plant materials is worked sometimes on a daily basis by means of a wide range of techniques. Despite this, plant processing still remains one of the most poorly understood human activities. Compared to other fields of research, even today, plant technology usually gets less attention then other topics.

We can see several reasons for this situation. Firstly, direct archaeological evidence of plant processing is rarely preserved on prehistoric settlements and that may have slowed the development of research. Secondly, few anthropologists and prehistorians are interested in plant technology as it represents activities which are not that spectacular, leaving behind only minor and discrete remains. Plant processing practices are generally dedicated to daily objects (basketry, net making, clothing, etc.) and performed with simple tools in the domestic sphere, while other, more spectacular and/or prestigious technologies such as metal working, monumental architecture, agriculture or hunting clearly got the best of our attention.

The present publication, which results from a round table held in Ghent in November 2006, brings together contributions from specialists with an archaeological or ethnographic background. All contributions focus on the technical aspects of plant processing either in a prehistoric or in an ethnographic perspective. They focus on techniques and use the concept of "chaîne opératoire". In other words, they study the entire sequence of processing, from raw material selection to manufacture, use and discard.

Five papers are dealing with plant processing in prehistoric contexts. Karen Hardy discusses plant data from European Late Palaeolithic and Mesolithic using ethnographic data from the sedentary Wola in Highland New Guinea. The production of bast fibres for string manufacture is one of the most important and time-consuming daily activities among the Wola.

Valérie Beugnier focuses on the results of microwear analysis on three sites belonging to the Early Mesolithic and the Final Neolithic from NW Belgium. On all three sites plant working is very dominant and the discussed data constitutes the earliest indications of plant processing among hunter-gatherers in Europe. Valérie Beugnier also tackles the problem of the use of serrated-edge tools, typical of various Neolithic cultures in NW Europe.

The same topic is discussed by Linda Hurcombe using data from extensive experiments focusing on the "chaînes opératoires" of strings made from tree bast and nettle bast fibres.

Two other papers – one by Emmanuelle Martial and Fabienne Médard and another by Susanna Harris – concern the production of textiles in Neolithic Europe. The first paper tries to correlate remains of textile production from some well-preserved Final Neolithic sites in northern France with specific archaeological features. The second paper mainly focuses on cloth evidence from the waterlogged site of Hornstaad in a spatial and social perspective.

The only purely anthropological paper is written by Damien Davy, who discusses the production of basketry based on a four years investigation of 18 socities in French Guyana. In this paper the focus is lying on the technological aspects but also on the cultural, social and symbolic meaning of basketry.

PRÉHISTOIRE ET ETHNOGRAPHIE DU TRAVAIL DES PLANTES

Depuis quelques années, différentes recherches nous ont permis de mesurer l'importance du travail des plantes dans les sociétés traditionnelles, préhistoriques et actuelles (Heider 1970, Pétrequin et Pétrequin 1988, Sillitoe 1988, Coles and Coles 1989, Owen 1993, Barber 1994, Adovasio et al. 2007). Dans ces communautés, la transformation des matières végétales occupe une place au sein de tous les secteurs économiques tels que l'alimentation, l'artisanat, l'armement, la pharmacopée, les rituels et l'art. Une grande diversité de produits, d'objets, d'armes et d'outils sont fabriqués et une grande variété de plantes sont collectées et traitées, selon des chaînes opératoires également diverses. Dans certains groupes, le travail des plantes apparaît, en outre, comme l'une des activités dominantes, en termes notamment de temps de travail consacré.

Ce domaine d'activité reste pourtant, à ce jour encore, peu étudié. En contexte archéologique, ce relatif désintérêt semble en partie dû à l'extrême pauvreté de la documentation disponible, les produits, objets et déchets végétaux n'étant, par nature et à de rares exceptions près, jamais conservés dans les sites.

Mais il semble aussi qu'un certain nombre de préjugés soit à l'origine de cette situation. Le travail des plantes est un artisanat généralement discret, peu spectaculaire et peu valorisé, plutôt lié à la sphère domestique. Dans la plupart des cas, les productions, réalisées dans le cadre d'activités quotidiennes, consistent en biens ou produits usuels et ne font pas l'objet d'échange, de transaction ou de don. Il s'agit, par exemple, de préparations alimentaires à base de végétaux ainsi que divers objets aussi banals que les cordes, les fils, les paniers, les filets de femmes, certains vêtements, etc. On peut ainsi opposer le travail des plantes à tout autre domaine technique beaucoup plus spectaculaire et valorisé, objet bien souvent aussi de plus d'attention de la part des chercheurs, tel que la chasse, l'agriculture, les constructions architecturales monumentales, la métallurgie, entre autres choses. Ce monde tend ainsi à disparaître, en ne laissant derrière lui que peu ou pas de traces, nous privant ainsi d'un aspect important de la vie quotidienne de ces sociétés.

Des recherches récentes existent toutefois qui montrent bien l'intérêt et la richesse de ce domaine et l'urgence qu'il y a à recueillir les derniers témoignages de pratiques artisanales encore peu connues. Pour la préhistoire, la documentation est également disponible mais reste fragmentaire et bien souvent difficile à exploiter et mettre en valeur.

Ce constat établi, la nécessité de réunir archéologues et ethnologues autour du thème du travail des plantes a émergé, le but de la rencontre étant d'obtenir un aperçu des différentes recherches menées actuellement sur la question. Il s'agissait également de pouvoir entamer une réflexion sur les difficultés communes à toutes ces études et d'illustrer concrètement quel avantage retirer de collaborations actives entre disciplines complémentaires. Une partie des contributions proposées lors de ces rencontres sont maintenant publiées, compte tenu de leur intérêt et de la cohérence qui est apparue entre chacune d'elles. À partir de la reconstitution des techniques liées au traitement des matières végétales, tous ces travaux abordent une même série de questions touchant au fonctionnement et à l'histoire des groupes étudiés.

Pour les périodes préhistoriques, cinq articles sont proposés. Karen Hardy réinterprète la documentation disponible concernant le Paléolithique supérieur et le Mésolithique européen à la lumière de données ethnographiques issues d'enquêtes menées en Papouasie-Nouvelle-Guinée. Celles-ci permettent de mettre en évidence l'importance des activités liées à la production de fils et de cordes en fibres végétales et les répercussions d'une telle production sur l'organisation de ces sociétés.

Valérie Beugnier présente les résultats d'analyses tracéologiques réalisées sur trois sites du Mésolithique ancien et du Néolithique final du nord de la Belgique, tous caractérisés par l'importance du travail des plantes. Ont ainsi été mis au jour les plus anciens témoins d'une exploitation massive des plantes à l'aide d'outils en silex en contexte, qui plus est, de campements de chasse. La question des microdenticulés et de leur énigmatique fonction est également largement abordée.

L'article de Linda Hurcombe traite également de la question des microdenticulés et de leur difficile interprétation fonctionnelle pour lesquels a été mis sur pied le plus important programme expérimental jamais réalisé, concernant le traitement des plantes et plus spécifiquement les chaînes opératoires de production de fils à partir des fibres d'orties et du liber d'écorces de tilleul et de saule.

Les articles d'Emmanuelle Martial et Fabienne Médard, d'une part, et de Susanna Harris, d'autre part, traitent quant à eux des productions textiles au Néolithique. Dans le premier, une approche originale est présentée fondée sur la collaboration entre plusieurs spécialistes et archéologues, de façon à pouvoir identifier les vestiges archéologiques liés à la fabrication des textiles. Dans le deuxième article, l'auteur aborde les notions de temps et d'espaces en rapport avec les chaînes opératoires de production des textiles.

Plus proche dans le temps, Damien Davy présente les résultats de quatre années d'enquête auprès de 18 communautés de Guyane française au sein desquelles les activités de fabrication des vanneries joue un rôle majeur. La place occupée par cet artisanat tant au niveau culturel, social que symbolique est discutée.

AUTHORS' ADDRESSES

Valérie Beugnier
Institut royal des Sciences naturelles de Belgique
Section anthropologie et Préhistoire
29 rue Vautier
B-1000 Bruxelles, Belgique
vbeugnier@yahoo.fr

Philippe Crombé
Ghent University
Department of Archaeology and Ancient History of Europe
Blandijnberg 2
B-9000 Gent, Belgium
philippe.crombe@Ugent.be

REFERENCES

ADOVASIO, J.M., SOFFER, O. and PAGE, J., 2007. *The Invisible Sex*. New York: Harper Collins, Smithsonian Books.

BARBER, E.W., 1994. *Women's Work: The first 20,000 years. Women, Cloth and Society in Early Times*. New York: W.W. Norton and Co.

COLES, B. and COLES, J., 1989. *People of the Wetlands. Bogs, Bodies and Lake-Dwellers*. London: Thames and Hudson.

HEIDER, K.G., 1970. *The Dugum Dani*. Chicago: Aldine Publishing Company.

OWEN, L.R., 1993. Material worked by hunter and gatherer groups of northern North America: implications for use-wear analysis. *In* : P. Anderson, S. Beyries, M. Otte and H. Plisson, eds. *Traces et fonction : les gestes retrouvés. Actes du colloque international de Liège, 8-10 Décembre 1990*. Liège : ERAUL 50, 3-12.

PÉTREQUIN, P. et PÉTREQUIN, A.-M., 1988. *Le Néolithique des lacs. Préhistoire des lacs de Chalain et de Clairvaux*. Paris : Éditions Errance.

Sillitoe, P., 1988. Made in Niugini. London: British Museum Publications.

WHERE WOULD WE BE WITHOUT STRING? ETHNOGRAPHIC AND PREHISTORIC EVIDENCE FOR THE USE, MANUFACTURE AND ROLE OF STRING IN THE UPPER PALAEOLITHIC AND MESOLITHIC OF NORTHERN EUROPE

Karen HARDY

Abstract: The role of string or cordage in prehistoric life is demonstrated both by bringing together the evidence from archaeological sites and also by exploring the manufacturing process of string and its central role in prehistoric life through use of ethnographic data from Papua New Guinea. Twisted bast fibre manufacture is long, slow work and the demand for string meant pressure of production was constant both with regards to the string itself and for artifacts such as bags and nets constructed solely out of string. Based on direct and indirect evidence for string in the Upper Palaeolithic and Mesolithic, the demand is likely to have been no less intense then than it was recently in highland Papua New Guinea. The implications of this for our understanding of Upper Palaeolithic and Mesolithic life are important as the time factor involved in this constant production may have been a defining aspect of life.

Keywords: string, cordage, bast fibre.

Résumé : Le rôle des fils et des cordes pendant la préhistoire est ici largement mis en évidence par l'analyse, d'une part, des témoins issus des sites archéologiques et, d'autre part, des données ethnographiques provenant de Papouasie-Nouvelle-Guinée. L'ethnographie permet notamment d'étudier la chaîne opératoire de fabrication du fil et son rôle central en préhistoire. En Papouasie-Nouvelle-Guinée, la fabrication de fil en fibres végétales est un travail long, fastidieux et mené sous la pression, de façon à répondre à une demande constante en ficelles et cordes, ces produits étant utilisés pour eux-mêmes mais aussi pour la fabrication de nombreux objets tels que les sacs et les filets. En s'appuyant sur les témoins directs et indirects disponibles dans les sites archéologiques, il semble que la demande en ficelles et cordes, au Paléolithique supérieur et au Mésolithique, n'ait pas été moins intense qu'il y a peu dans la région des Hautes terres de Papouasie-Nouvelle-Guinée. Cette donnée est fondamentale pour notre compréhension de ces périodes de la préhistoire, le temps consacré à cette production incessante ayant pu être un aspect déterminant de la vie quotidienne.

Mots-clés : fil, corde, fibre de liber (filasse).

INTRODUCTION

« The Art of Ropemaking, by some strange fatality, has not attracted hitherto sufficiently the notice or attention of the mathematician, philosopher, or engineer, either in this country, or any part of the maritime world, with success » (Chapman 1868).

Though the above comment was written in reference to more recent historical rope-making it could equally well apply to the interest afforded the prehistory of string. In the last few years, a small but dogged group of researchers have worked to promote their vision of the significance of the « extremely unsexy » (Adovasio *et al.* 2007, chap. XII) world of perishable prehistoric artifacts. That this world can be described as unsexy is testament less to its lack of sex appeal and more to the almost generalized ignorance about the crucial importance and role of twisted fibres and artifacts made from these, to us and to our past.

To understand the importance of string, cordage or the metal version, wire, in our lives today is very simple. Virtually all items of clothing, except those made from polyester, are based on woven cloth, the cloth itself spun from natural plant fibres, electricity and telephones are based on wires, suspension bridges, cranes, stringed musical instruments, sailing ships, shoe laces and so on are all based on the technology of lines. Cord-less phones and wire-less internet are two of the most recent technologies to have been invented. There is even a String Theory to explain the origins of the universe while string or something that ties things together is considered to be one of only a handful of cultural universals, that is, something that every known human group has (Brown 1991).

Awareness of the crucial importance of fibre technology in the past is not new (Warner and Bednarik 1996, Good 2001) but is not widely known. E. W. Barber (1994) describes it as « the unseen weapon that allowed the human race to conquer the earth » while J. M. Adovasio *et al.* (2007) point out that where these survive in archaeological contexts such as dry caves, fibre artefacts outnumber stone tools by a factor of 20 to 1 while in anaerobic conditions 95 % of all artefacts are either made of wood or fibre. This is unsurprising when one considers the range of artefacts that must have used string as a fundamental component part, clothing, bags, nets and virtually all other fishing gear, bows and arrows, harpoons, containers, footwear, carrying gear, lashings for houses, and boats, jewellery and personal adornment, in effect, everything except perhaps babies nappies. String was undoubtedly one of the key technologies of the Palaeolithic and Mesolithic and was probably no less significant than it was to the Dani of highland New Guinea. « The Dani are technically a Stone Age culture. Stone tools are important but in fact the Dani culture is based on wood and string and could be called a String Culture » (Heider 1970).

All the available evidence points to tree bast as the earliest raw material used to twist fibres in the Eurasian Palaeolithic and Mesolithic and the focus of this paper is on functional and social aspects of twisted fibre string.

EARLY EVIDENCE FOR STRING

The knotting of fibres may have a very long history indeed. Gorillas and chimps manipulate fibres and make knots. C. Warner and R. Bednarik (1996) suggest that this could indicate an ancestry for the early use of knot tying dating to before the evolution of the genus *Homo*.

The development of composite technology, that is anything that is « made up of disparate or separate parts or elements » (Collins English Dictionary 1999) was enabled largely through the use of string or a natural alternative such as vine. The only other way in which two objects could be attached was with glue.

Evidence for the use of birch bark tar as glue, appears earlier in the archaeological record, during the Middle Pleistocene adhering to two flint flakes, one of which was still hafted (Mazza *et al.* 2006). Two pieces of birch bark pitch, one of

which contained a fingerprint impression and the imprint of a stone tool and the structure of wood cells (Koller *et al.* 2001) were found near Konigsaue in Germany dating to the Middle Palaeolithic aged somewhere between 40–80,000 years old. These pieces have been interpreted by the authors as hafting glue for attaching a flint tool onto a wooden haft.

Woodworking undoubtedly also has a very long history. C. Warner and R. Bednarik (1996) claim the earliest evidence for woodworking may be the use of boats, by 700,000 years humans had crossed onto the island of Flores. However they also describe other means of getting about in the water such as using inflated animal skins and reed rafts. Additionally, B. Malinowski (1922) describes the construction of an ocean going canoe in which the lashing holding all the bits together is a vine. It is therefore conceivable that sea-going craft could have been manufactured using neither woodworking nor manufactured cordage. Evidence for sea crossing should not therefore be considered alone as absolute evidence for early woodworking or for manufactured string. M. Dominguez-Rodrigo *et al.* (2001) claim evidence for woodworking based on the detection of adhering phytoliths on Acheulian tools dated to between 1,7–1,5 million years while M. Lillie *et al.* (2005) highlight evidence of woodworking in the Mesolithic.

The second great invention based on string was the development of technologies such as looping and weaving that enabled artefacts made from woven materials to develop (Barber 1994). These include items such as bags and nets. This is likely to have revolutionised hunting, fishing and small item collecting (e.g. shellfish, berries). J. M. Adovasio *et al.* (2007) suggest the use of nets in hunting as well as for fishing.

Raw materials used for making string include fibrous plants such as nettle, as well as sinews and hide; this is particularly common in extreme environments such as the Kalahari (Lee 1979) and the Arctic (e.g. Garth Taylor 1974) though even here it is common to find both animal and plant raw materials used to make string.

In archaeological contexts, possibly the earliest concrete evidence for the use of cordage comes from Repolusthöhle in Austria; here two perforated objects, a wolf incisor and a bone point, were found in a context which is understood to date to 300,000 years old, making them the oldest perforated artefacts in the world (Bednarik 1995, Warner and Bednarik 1996). If indeed they were pendants, then string of some sort would have been required to hang them. R. Bednarik (1997, 2000) describes ostrich eggshell beads that date to around 200,000 years from Libya, « the earliest known disc beads in the world » (Bednarik 2000) and highlights the presence of beads of great antiquity from many places across the world including Africa, India, China, Mongolia, Russia and Siberia. M. Vanhaeren *et al.* (2006) record perforated shells which they believe are beads dating to between 100,000 and 125,000 BP. C. Henshilwood *et al.* (2004) conducted a usewear study on apparently artificially perforated shells which they believe are personal ornaments, from the South African site of Blombos dated to between 75,000-80,000 BP. They found a use-wear pattern « consistent with friction from rubbing against thread, clothes, or other beads ». By 40,000 years ago, shell beads occur in large numbers on many sites, for example Ucagızlı Cave in Turkey (Kuhn *et al.* 2001) and Ksar 'Akil in Lebanon, and ostrich shell beads are found in many MSA sites in east Africa from 40,000 years ago (Ambrose 1998) and earlier (Bednarik 2000).

Other uses have been recorded for shells (e.g. Claassen 1998, Balme and Morse 2006) though these do not need to be perforated if they are not going to be strung or tied. Ethnographic evidence points to a variety of uses for perforated shells. Cowries for example are used for decoration (e.g. Jackson 1917, Sillitoe 1988, Carey 1998), as currency, gift exchange, in medicine (particularly against smallpox), to convey messages or ideas in code, as charms, as net sinkers, for divination, and in China they were used with rice to stuff the mouths of the dead (Jackson 1917, Mair 1969, Claassen 1998, Sciama and Eicher 1998, Gabiole 2004). If these early examples of perforated shells are indeed personal ornamentation as is suggested in particular by the use wear work described in Henshilwood *et al.* (2004) they are likely to have been either attached to clothing or strung as neck-laces, bracelets or anklets. Fine thread is required to attach beads to clothing and string of some sort is required to make necklaces or bracelets (fig. 1). These pendants and beads therefore may represent the earliest indirect evidence for the use of string or cordage in the archaeological record.

A more direct piece of evidence for the use of twisted fibres dates to 27,000 BP. O. Soffer (2004, Soffer *et al.* 2000) and E.W. Barber (1994) record impressions in clay of imprints of complex items of woven material, from « dressed » figurines. O. Soffer (2004, Soffer *et al.* 2001) also records the presence of eyed needles from this time.

Though the specific plant types of the earliest finds of twisted fibre are not identified as to

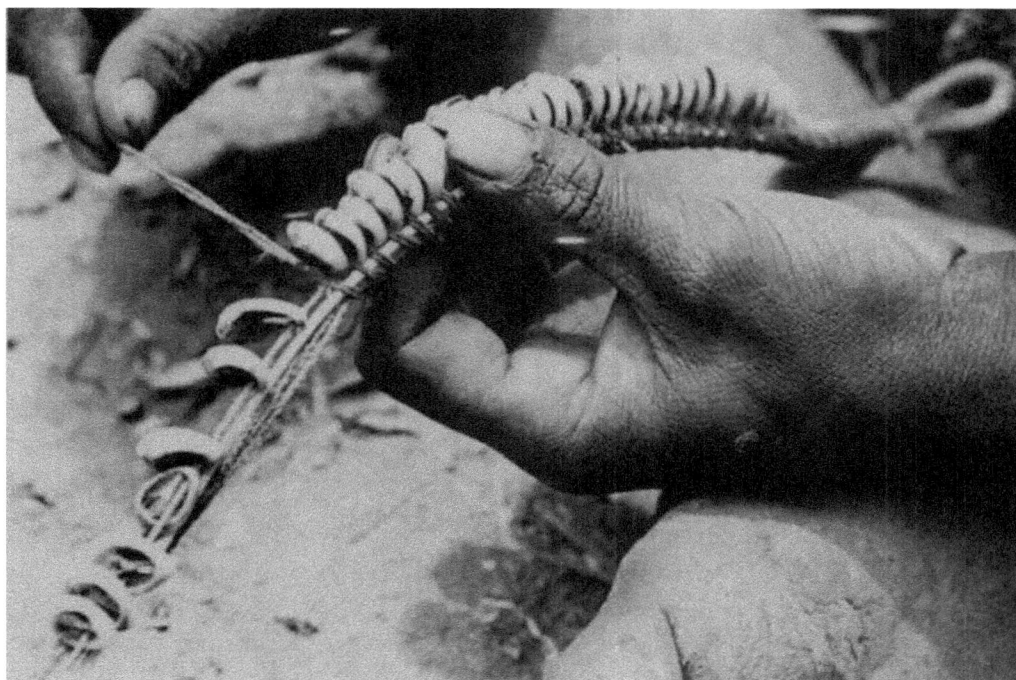

Figure 1. Attaching cowrie shells to a side mounted necklace in Papua New Guinea (Hardy and Sillitoe 2003).

species, the woven material on the « dressed » figurines was identified visually by O. Soffer (2004) as bast fibre. Therefore by early in the 30th millennium BP twisted fibre technology had been mastered to a level that enabled complex manipulation of the string to form complete artefacts such as items of clothing through looping and knotting. To have reached this level of sophistication at this stage of the Upper Palaeolithic suggests twisted fibre technology has a very deep ancestry.

The oldest actual pieces of twisted and plied fibre are three fragments found at Ohalo II in Israel dating to 19,000 BP (Nadel *et al.* 1994). More twisted fibre fragments come from Lascaux and date to 17,000 years (Leroi-Gourhan 1982).

Evidence for the use of twisted plant fibres in the northern European Mesolithic comes form a variety of sites, many around the Baltic region.

The earliest physical evidence for the use of plant fibres in northern Europe comes from Friesack in Germany where many fragments of nets, ropes and thousands of fragments of twisted bast fibre string were found which date to the early part of the 10th millennium BP (Gramsch 1992). Additionally several bone points had narrow strips of bast fibre twisted around them, clearly the remains of some type of hafting. Both knotted and knotless netting (nålebinding) are represented, using twisted bast fibre. In Denmark, a number of sites have also produced fragments of textiles and string made from plant material. Sites include Tybrind Vig (Andersen 1985) where textiles, string and thread were found, some of which were identified possibly as willow bast (Bennike *et al.* 1986); Kongsted Lyng and Tulstrup Mose where coarse fibre was identified as lime bast and worked using twined weave (Becker 1947); Ulkestrup Mose where a bone point was found with the shaft attached by twisted bast fibre, most likely lime; and the early Neolithic site of Bolkilde, where three textile fragments woven using the knotless netting (nålebinding) technique and a plaited piece were found (Bennike *et al.*1986). At Bolkilde the textile remains lay beside the skeleton which suggests that they might have been clothing.

Several other Danish Mesolithic and early Neolithic sites have produced evidence of twisted plant fibre string. These include Skoldnaes, Derjø (Skaarup 1982), Møllegabet II (Skaarup and Grøn 2004) and Sigersdal Mose (Bender Jørgensen 1986). A fragment of twisted bast fibre net was found at the Mesolithic site of Antrea in Finland.

This fragment was identified as willow bast and it was knotted (Äyräpää 1950). Another fragment of net came from the Estonian Mesolithic site of Siiversti. Again this was made from twisted bast fibre, though this sample had been woven using the knotless netting or nålebinding technique (Clark 1952). At the Mesolithic site of Forster-moor in Germany bast string was found on a small bow, while at Vinkelmore and Fippenborg, both Mesolithic sites, string was found wrapped round arrows, perhaps to hold feathers. Other sites also have remains of string but it is not clear whether these are made of twisted bast fibre or some other material (Mertens 2000).

Numerous twisted plant fibre fragments were found at Vis I in Russia where they were knotted (Burov 1998). Many Neolithic sites have also produced large numbers of twisted inner bast fragments; for example at Sārnate in Latvia the remains of a net was found as well as stone sinkers and floats with bast fibre wrapped round them, a fragment of knotted netting was found at Abora I Latvia, (Bērzinš 2006), a net and a household mat made of twisted linden bast was found at Šventoji, Lithuania (Rimantiené 2005) and while many other sites across northern Europe and Russia, Estonia, Latvia and Lithuania also contained fragments of twisted bast fibre or netting. (e.g. Vogt 1937, Burov 1967, Indreko 1967). It appears that a range of basts from different trees notably willow, lime and oak were used to make fibres at this time though it is not clear whether the different basts were used to make string or cordage for different purposes. Ethnographic evidence (Sillitoe 1988) suggests the selection of specific fibres for specific purposes on the basis of strength, colour and coarseness.

In addition to the finds of actual cordage, there is a range of secondary evidence for string from Mesolithic sites in the form of fishing gear, nets, traps, harpoons, needles, or bodkins and perforated shell beads (Albrethsen and Brinch Petersen 1976, Mellars 1987, Mordant and Mordant 1992, Soffer 2004, Hardy in press a, b). Even when these items are missing, indirect evidence such as, for example, use of a catch-all method of fishing, demonstrates the use of nets or traps (Parks and Barrett in press), the secondary evidence for carrying gear (Hardy forthcoming) in the form of shell middens (e.g. Mellars 1987, Hardy and Wickham-Jones in press), caches of hazelnuts (Mithen 2000) and consumption of elm seeds (Grøn 1998). String bags lined with grass or moss can be used for gathering small items (e.g.

Harrington 1924) though other raw materials such as birch bark and hide could also be used to make containers (e.g. Lee 1979, O. Grøn pers. comm.).

The summary above demonstrates that the use of twisted bast fibre to make a range of artefacts is likely to have been widespread across Mesolithic northern Europe with twisted bast fibre occurring on many waterlogged sites, where the survival of organic remains is favourable.

PRODUCTION METHOD AND USE OF STRING FROM TWISTED FIBRES

Tree bast consists of fibres attached to the inner side of bark from stems of some dicotyledonous plants. The use of tree bast to make string and clothing is widespread in the ethnographic record and indeed bast fibre is still used widely today to manufacture cordage, from plants such as jute, hemp and flax. Bast has been used continuously at least since the Mesolithic (Myking *et al.* 2005)

and it has a wide range of uses. For example in Russia and Eastern Europe it was used to make shoes well into the 20th Century, while the production of bark fibre clothing continues in South East Asia (Howard 2006).

The manufacturing sequence of making string from bast fibre remains, even though it is now industralized, in essence the same as when it was first developed. A series of ordered steps have to be followed both in order to understand the earliest development of the technology and to practice it. This is sufficiently complex that it is likely to have required an extended period of time to develop.

Steps required to manufacture string from bast fibres and then to make woven or looped materials include:

1. Understanding the concept of fibres as a mechanism for attaching distinct objects.

2. Identification and selection of appropriate raw materials.

3. Extraction and preparation of fibres.

Figure 2. Stripping bark off saplings to make string in Papua New Guinea (Hardy and Sillitoe 2003).

Figure 3. Separating bast fibre from bark for making string (Hardy and Sillitoe 2003).

4. Joining and adding fibres through rolling or twisting to create extended lengths.

5. Adding in extra fibres in reverse twist to give extra strength (plying).

6. Weaving, looping and other complex technologies.

In the New Guinea highlands, non industrial manufacture of string from bast continues (e.g. Sillitoe 1988, Mackenzie 1991, Hampton 1999).

The New Guinea highlands comprise a series of mountain ranges occupying the interior of the island. They are physically isolated from the coastal region and were thought by outsiders to be uninhabited until the 1930s. Highland New Guinea is divided into many territories identified with small kin-constituted communities with average populations of around 300 people. They live in squat houses scattered along the valley sides, largely in areas of secondary regrowth. Homesteads can be of variable size and composition, comprising diversely made up family groups; men customarily occupy houses separate from women. They cultivate gardens in both grassland and forest, following a sedentary version of swidden cultivation and the neat gardens are dotted about the valleys. The sweet potato is the staple crop supplemented by a variety of other crops such as taro, banana, sugar cane and a variety of green vegetables (Bourke *et al.* 1995).

A total of 1035 plant species were known to be used still in 1976, most of them in construction of the material culture while 46 different plants were still being used to make string (Paijmans 1976). Different types of string or cordage required different plants, for example heavy cordage for animal tethers used bigger coarser fibres with thicker bunches rolled together. Other than this, there is no record for any specific selection of a particular tree or plant type for a particular type of fibre, however among the Wola for example, eleven trees were used, all of which produce fibres with individual qualities of strength, elasticity, pliancy and colour (Sillitoe 1988).

String making is based on extraction of fibres from bast, the inner part of tree bark (fig. 2 and 3). Bast is the food conducting tissue of vascular plants and consists of sieve tubes, fibres, parenchyma and sclerids. While string making will differ in a minor way according to climate and tree type if bast is used there is a general methodology that needs to be followed to extract and prepare fibres for use.

To obtain fibres raw materials have to be collected or traded. The Telefol (Mackenzie 1991) sometimes trade for better quality fibres rather than using their local raw materials, however the Wola collect all their own materials (Sillitoe 1988). The following description comes from M. Mackenzie (*op. cit.*) where a full account can be found. Once the bark is collected, it is cut by men then immersed in water or laid across the glowing embers of a dying fire to produce a gentle heat. The outer bark is then peeled off and placed on a smoking rack. Depending on the type of material it can take anything from one night to two weeks to dry out correctly. The flattened out fibre is then beaten and folded and rebeaten between flat river pebbles until the bast fibres are loosened and the fibre flattens out. This is then left to dry. Once dry it can be folded and stored. Both men and women participate more or less equally in this first stage. All the remaining parts of the process are carried out exclusively by women.

When the bark is needed, it is remoistened either by leaving it outside for dew moistening overnight, by chewing or by sealing in bamboo tubes. The women like to make string in the early morning while baking the breakfast when their fibres are full of dew. They say this makes the fibres easier to work. Once the fibres are ready, the first task is to shred them. M. Mackenzie (*op. cit.*) explains this is often done by young children.

Then the rolling begins (fig. 4). Women select fibres and being rolling them against the thigh to twist the fibres together, adding in new fibres where necessary. To extend the length of the string, fibres are added into it. To obtain a high quality string, it is important to have no bumps where the ends join up. The secret of this is to know the exact amount of new fibre to add in. The yarn is then plied in the opposite direction.

SOCIAL ASPECTS OF STRING MANUFACTURE

M. Mackenzie (1991) explains that children are experimenting with rolling fibres on their legs by aged two and by the age of seven, all girls have made their first looped bag and string making becomes integrated into their life.

They learn through copying and imitating rather than instruction to the point where, by adolescen-

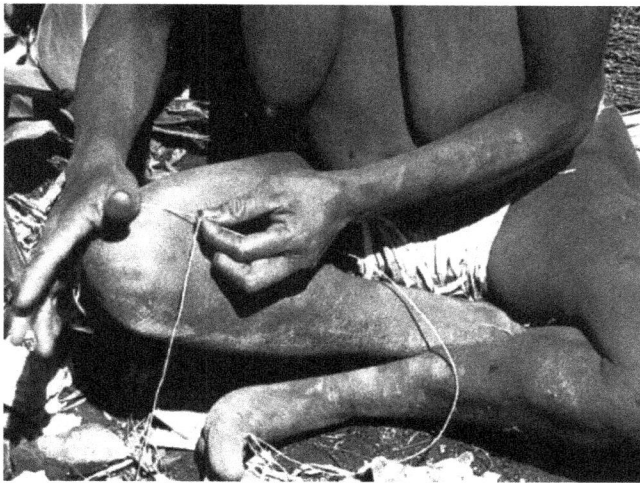
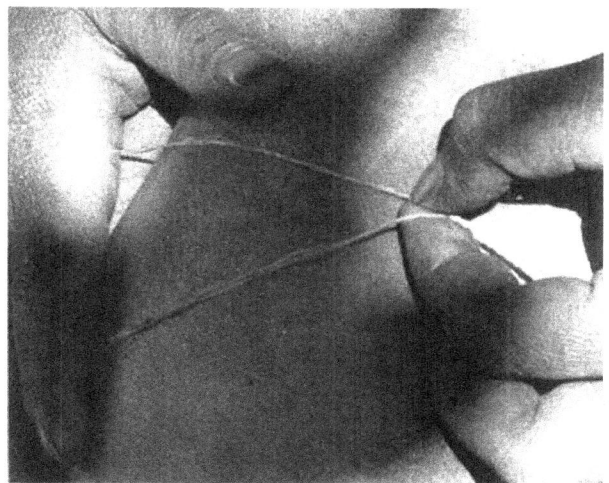

a   b

Figure 4. Sting making in Papua New Guinea:
a. Twisting the strands of string; b. Close-up (Hardy and Sillitoe 2003).

ce, string making becomes an automatic activity. This means that their actions become an integral part of their physical being as they are learned as the body is still growing (Hardy in press c). String making and looping therefore become automatic repetitive activities, not requiring conscious thought which enables the workers to carry out other activities, at the same time.

String, made from the ficus plant in the way described above, falls into average lengths of 4-5 metres which is about 20 reels across the hand. This is the ideal length for subsequent looping (Mackenzie 1991). Time estimates for raw material preparation and string production for 86-90 lengths of 4-5 metres, based on the ficus plant, is found in figure 5. This produces the average amount of string required for one bag. Beyond that is the time taken to do the looping to make the bag, which for this bag here was estimated as between an additional 100-160 hours. In addition to bags a wide range of other items are made either by looping string or by incorporating string into a composite artefact. When one considers the important functional and social roles of the bags and the fact that a woman might wear up to 7 bags

at any one time, the extent of string production requirements becomes apparent, it is no wonder that string production is a never ending process.

In order to gauge the demand for string in one society, a count of material culture items that used string as a component part was made. This information was taken from P. Sillitoe (1988). The material culture of the Wola of highland New Guinea consists of almost 200 individually named items. Of these, 72 % (144 items) had string as a component part. This does not include the use of rattan or vine or untwisted bast fibre.

Items that needed string to construct them included animal tethers, bows and arrows, fishing nets, all clothes, all wigs, all body decoration items and most musical instruments (fig. 6). If unrolled fibres, strands of vine and rattan are included, then it reaches around 95 % of items. In effect virtually no items, except items such as unmounted flaked artefacts and face paints, were constructed without the use of string or cordage of some type. P. Sillitoe (*op. cit.*) suggests that for average use, bags made from string last around 1-2 years before wearing out.

Raw material collection, separating the bast and drying the fibres	About 14 hours over a 2 week period.
Rolling the lengths 25-30 minutes per length	Between 65-80 hours
Total time	Between 79-94 hours

Figure 5. Time estimates for raw material preparation and string production for 86-90 lengths of 4-5 metres, made from the ficus plant. This is the required amount of string for one bag.

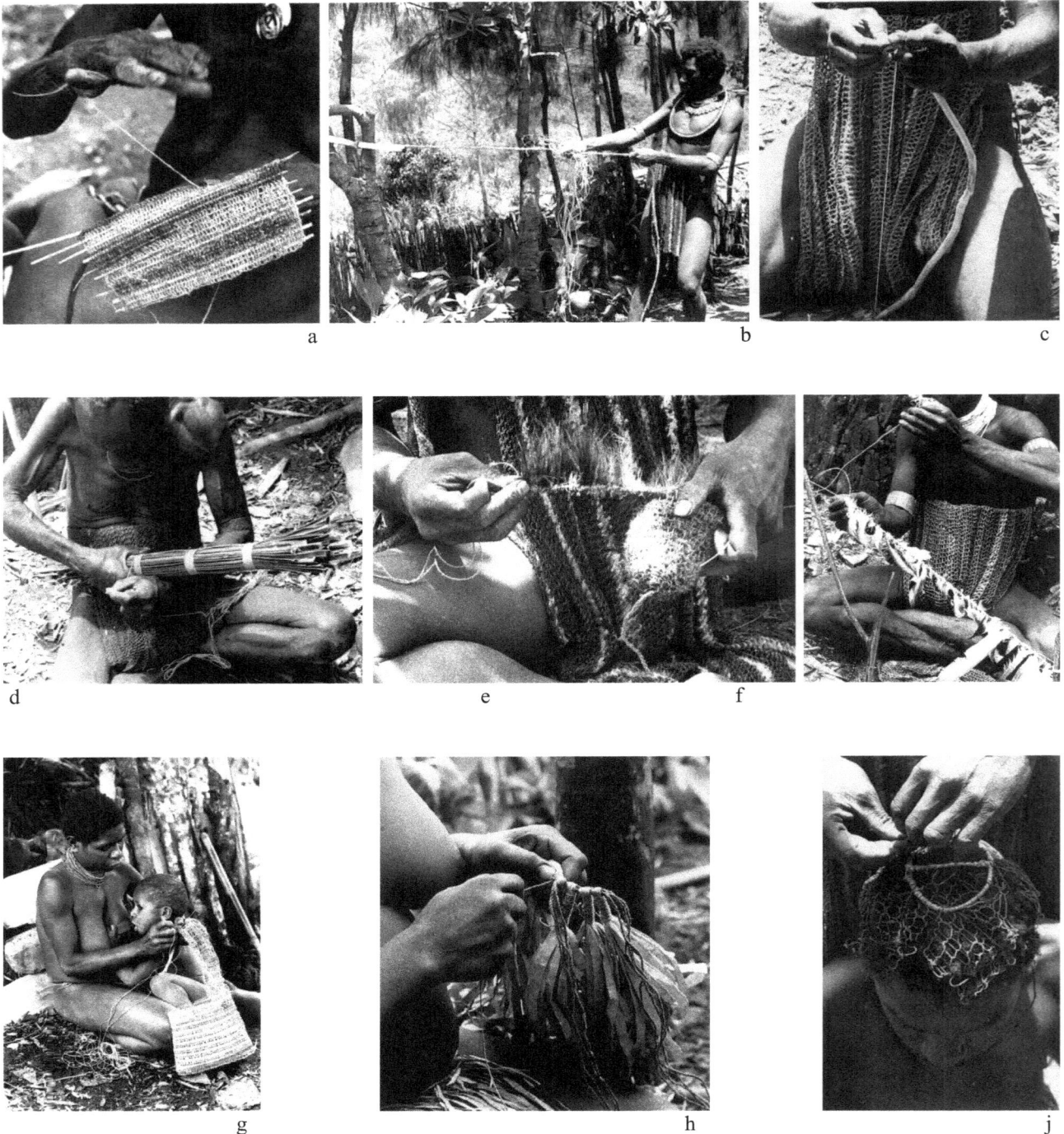

Figure 6. Examples of the use of string in Papua New Guinea:
a. Netting the crown of a fancy net-cape; b. Stretching and rubbing a pig tether; c. Stringing a mouth bow;
d. Winding string around the end of a cylindrical pendant; e. Coiling and stitching a pompon;
f. Securing feathers to a standing line for a feather circlet; g. Netting the strap of a man's wide-strap bag;
h. Whipping stout bundles of sedge onto the rear of a skirt; j. Positioning *kwiysh mor* crown support
on the head before building a burr wig (Hardy and Sillitoe 2003).

All string except animal cordage was made by women among the Wola. With so many items requiring string in their manufacture, string making was an everyday task. Using figures from Sillitoe (*op. cit.*), an estimate of 75 % of manu-facturing time was taken up with string making and looping items out of string. String making alone took up well over 50 % of women's manu-facturing time and 45 % of all manufacturing time when men's time was included also.

The consumption of string by the Wola people was ongoing and there was a constant demand for string. The pressure was on women to produce string. As string making was a uniquely feminine expertise, it could in theory have provided women with some sort of powerbase. No mention of this is made by P. Sillitoe, however the following quotation, from M. Mackenzie (1991), suggests that women were indeed able to exercise a certain degree autonomy at least over their own time, based on their unique manufacturing knowledge. « Men sometimes tell us to put our bilums away and cook food, or they talk cross because we are sitting making bilums, not working in the garden. But … we tell them, 'Making bilums is number one work. We can't go to the garden if we haven't got a bilum to fill up' » (Mackenzie 1991).

The manufacture of string from fibrous plants or bast is time consuming repetitive work. Indeed string manufacture may well have been the single most time consuming activity undertaken by the manufacturing group throughout their lives as they were completely tied to the ongoing demand for string and the need to produce. While string rolling can only be carried out sitting still, looping is an activity that can be performed while walking around and this is in effect what the women do, looping while walking to and from their gardens. This behaviour is little different from the way in which these skills were practiced elsewhere.

« Women who ran the croft and home, knitted whenever time allowed. It was not an uncommon sight to see a woman knitting as she carried peats in a « kishie » or basket from the peat bank to her croft house » (http://www.shetlandknitwear.com/history.html).

It is of course impossible to determine whether the gender specializations apparent in highland New Guinea, where women make virtually all the string and all the items manufactured uniquely from string, have any connection with those in prehistory. Certainly among the Wola, there is a clear distinction between working with « strong » raw materials, (bone, wood, etc.) which is men's work, and « soft » ones mostly plant material which is women's work (Hardy and Sillitoe 2003, Sillitoe and Hardy 2003). This clear gender-based distinction between strong and soft raw materials which is described as such by the Wola is interesting, and the use of strong and soft is extended to cover other aspects of the distinction between the roles of men and women that the Wola employ (Sillitoe 1988).

While it is not possible to be sure that it was women working with plant fibres in Mesolithic Europe, there is widespread ethnographic evidence from across the world to suggest that women are more closely linked to « soft » plant materials. Additionally, though perhaps a tenuous link, the fact that women across Europe still now work with their hands doing automatic repetitive tasks with thread, wool or fibres might be significant. Examples include knitters from British islands, Flemish lace makers, Romanian spinners and Bulgarian knitters.

Nålebinding (knotless knitting or needle binding) survives in several geographically separate locations including remote parts of Central Asia, northern Scandinavia and New Guinea (Decker 2000), and in these places it is almost exclusively employed by women. Research into reasons why women appear to have better fine motor, spatial location and object memory skills, have suggested that these abilities are all thought to have developed out of an early focus on plant detection (Kimura 1996).

STRING AND MAGIC

Among the Telefol women of highland New Guinea left their bast outside in the dew overnight, ostensibly to soften and dampen it. However dew is also considered by the Telefol as a symbol of increase and growth. Men for example rub themselves with dew in the morning and the women may have used the dew as a type of magical component in their string making to promote strong and sturdy string (Mackenzie 1991).

An insight into the importance of string or tying fibre to prehistoric people is provided by B. Malinowski (1922). In his account of the construction of an ocean going canoe, it becomes clear that the lashing, or cordage is clearly the most important part of the boat.

Stage one is the selection and felling of the tree, digging out the trunk, and preparing the planks, boards, poles and sticks. This can take up to 6 months and is done by the primary canoe builder along with a few helpers.

The second stage is communal work and takes only a few days. This stage involves piecing together all the planks and prow boards and getting them all to fit exactly. Then they are lashed together.

Following this, the outrigger is lashed, the canoe is painted and the sails are made. These stages of canoe building are accompanied by magic and exorcisms. The lashing of the canoe is accompanied by the most important magic of all. The material used in lashing is a creeper called wayugo.

Wayugo is the only material they will use for the lashing and it is considered of the utmost importance that this creeper is sound and strong as it is this alone which maintains the cohesion of all the parts and keep the canoe together. In rough weather particularly, the single most important issue is that the lashings are strong enough to withstand the strain or the waves, their lives depend completely on the soundness and strength of their lashing. The wood can be tested in an ongoing fashion but the element of danger and uncertainty in a canoe is due mainly to the lashings.

It is for this reason that the magic of the lashing is one of the most important rituals in canoe construction to the extent that wayugo, the name of the creeper, also means canoe magic. When a man has a reputation for building a good or fast canoe he is described as having or knowing a good wayugo.

Other examples of the life dependence on rope or string include the ropes used to descend cliffs to collect sea-birds on St Kilda (Steel 1994) and the seal-hide thongs used by the Ona of Tierra del Fuego to lower themselves down cliffs to catch sleeping birds (Bridges 1951).

DISCUSSION

It is not easy to obtain direct evidence for the manufacture of string in the Mesolithic. However thanks to the few lucky survivals like those from Denmark, and the exceptional finds of impressions of netting in clay, we know that string was not only being made from the inner bark of trees or bast in the Mesolithic but that it was also being woven not only in the Mesolithic but also well into the Palaeolithic. However to expand our understanding of the implications of string making into the lives and material culture of the inhabitants of Mesolithic Europe, we need to turn elsewhere for help.

Ethnographic evidence needs to be processed with caution but it can provide an insight into a world that we normally cannot even imagine and one which, when the presence of non native raw materials does not intrude too deeply, can give us an idea of the scope and extent of the use of plant fibres, something that is also highlighted by Adovasio's percentages of 80 to 95 % percentages for artefacts of organic materials found on some archaeological sites.

Another way forward for detection of plant use is lithic use wear analysis based on experimentation, though in reality this tells us little more than that a particular type of lithic artifact may have been used in plant processing.

The use of residues, in particular starch granules and phytoliths, both of which can be taken from artefacts and soil profiles is an exciting method that allows us to reconstruct, in a direct and empirical way, the presence of certain plant types in a given locality. This is already proving successful for example the presence of a brush fence was indicated at the Neanderthal site of Tor Faraj through distribution of phytoliths (Henry et al. 2004). Several species of tree such as elm and birch for example, store their starch in the inner bark which is also edible. Use of inner bark may therefore become apparent through the identification of starches.

The ongoing demand for string and the probable lifelong obligations to make string will have been a major factor in the lives of the manufacturers. String manufacture is likely to have been one of the most time-consuming aspects of their lives and our image of them should include a bundle of string and busy hands as they went about their business. Indeed the construction of the technological items required to obtain food is likely to have been as significant in terms of ongoing daily obligations as the food collecting activities themselves. If indeed it was women who were the primary makers and loopers of string then our image of them should include bundles of fibres and busy hands wherever they were and whatever else they happened to be doing.

The way in which the bast fibres had to be prepared for use must have had an impact on the patterns of movement of nomadic people. The people from highland New Guinea, who are sedentary, had little restraint on the time needed to prepare the fibres for string making, however a mobile hunter gatherer group had to stay in one place for sufficient time to prepare the fibres and have completed the fibre preparation and string manufacture before they moved. Though movement has largely been defined by archaeo-

18

logists on the basis of availability of food resources, the reality is likely to have been much more complex.

In Patagonia for example movement was as likely to occur for social or other non food related reasons, such as for example because all the firewood nearby has been used up. Maritime Mesolithic Northern Europe, will have had a very rich resource base (e.g. Hardy 2007), but in order to access it, an array of material culture items, for catching fish, for collecting small items, for hunting animals, for digging up tuberous plants and so on, would have been essential and all of them needed to be made, something that involved collection of raw materials, preparation of raw materials and construction of the artifacts themselves. Many of these items will have involved « soft » or plant materials, most of which would not survive in an archaeological context. The examples which do survive, mainly from Baltic sites, give us an idea of some of the items used.

With very few exceptions (Clark 1976, Zvelebil 1994) plants have largely been ignored in our attempts to understand Palaeolithic and Mesolithic life from the perspectives of both material culture and food. Ethnographic information has repeatedly demonstrated that indigenous knowledge of plants is extensive (e.g. Owen 1993), and this is confirmed by the evidence in the form of artifacts requiring string and the actual remains of twisted fibres that have been found on certain sites with exceptional preservation as well as the ethnographic record from across the world.

People look at Europe today and see a bare and treeless landscape. However during the Mesolithic this would not have been the case. Northern Europe would have been densely packed with enough plants to make as much string as required and as many elaborate items of material culture as anyone could invent.

Throughout the world non industrialised people know their plants, they know their names and they know their uses, both as food and for other purposes. Extensive empirical evidence for the use of twisted fibres in the Palaeolithic and Mesolithic is unlikely, but by extending the lines of inquiry, the presence of twisted fibre technology and the intrinsic role of string begin to become apparent.

So much has been learned about the Mesolithic in the recent past. But plants, possibly the major factor both in diet and material culture, is still largely missing from our understanding of the past. Plants are centrally important to human groups across the world and they are likely always to have been. There is enough direct evidence available now to tell us this, from sites like Tybrind Vig, both in terms of food and as raw material for material culture, yet the role of plants in the lives and diet of people is still not included as a matter of course because direct evidence in the form of plant remains that can be seen and counted are normally not found. It is time that this changed and plants are afforded their rightful place in our past.

ACKNOWLEDGEMENTS

Thanks to Ole Grøn, Flora Gröning and Ulrike Sommer for help with the German texts and also to Ilga Zagorska for information and texts on material from the eastern Baltic. The author was in receipt of an EU Marie Curie Outgoing International Fellowship while researching and writing this paper. All photographs were taken by Paul Sillitoe.

AUTHOR'S ADRESS

Karen HARDY
BioArch
Department of Biology
P. O. Box 373
University of York.
York YO 10 5YW - UK
karhardy@gmail.com

REFERENCES

ADOVASIO, J. M., SOFFER, O. and PAGE, J., 2007. *The Invisible Sex*. New York: Harper Collins, Smithsonian Books.

ALBRETHSEN, S. E. and BRINCH PETERSEN, E., 1976. Excavation of a Mesolithic cemetery at Vedbeck, Denmark. *Acta Archaeologica,* 47, 1-28.

AMBROSE, S. H., 1998. Chronology of the Later Stone Age and Food Production in East Africa. *Journal of Archaeological Science,* 25, 377-392.

ANDERSEN, S. H., 1985. Tybring Vig. A preliminary report on a submerged Ertebølle settlement. *Journal of Danish Archaeology*, 4, 53-69.

ÄYRÄPÄÄ, A., 1950. Die ältesten steinzeitlichen Funde aus Finnland. *Acta Archaeologica*, 21, 1-43.

BALME, J. and MORSE, K., 2006. Shell beads and social behaviour in Pleistocene Australia. *Antiquity*, 80 (310), 799–811.

BARBER, E. W., 1994. *Women's Work: The first*

20,000 years. *Women, Cloth and Society in Early Times*. New York: W. W. Norton and Co.

BECKER, C. J., 1947. *Mosefundne lerkar fra yngre Stenalder*. Aarbøger for nordisk Oldkyndighed og Histoire.

BEDNARIK, R. G., 1995. Concept-mediated marking in the Lower Palaeolithic. *Current Anthropology*, 36 (4), 605-634.

BEDNARIK, R. G., 1997. The role of Pleistocene beads in documenting hominid cognition. *Rock Art Research*, 14, 27-41.

BEDNARIK, R., 2000. *Beads and the origins of symbolism. Semiotica Home Page*. Available from: www.semioticon.com/.

BENDER JØRGENSEN, L., 1986. The string from Sigersdal Mose. *Journal of Danish Archaeology*, 5.

BENNIKE, P., EBBESEN, K. and BENDER JØRGENSEN, L., 1986. Early neolithic skeletons from Bolkilde bog, Denmark. *Antiquity*, 60 (230),199-209.

BĒRZIŅŠ, V., 2005. Net fishing gear from Sārnate Neolithic site, Latvia. *In: People, Material Culture and Environment in the North. Proceedings of the 22nd Nordic Archaeological Conference, University of Oulu, 18-23 August 2004*. Riga: Latvijas Vēstures Institūta Apgāds, 37-61.

BOUEKE, R. M., ALLEN, B. J., HIDE, R. L., FRITSCH, D., GRAN, R., HOBSBAWM, P., KANABE, B., LEVETT, M. P., LYON, S. and VARVALIU, A., 1995. *Southern Highlands Province: text, summaries, maps, code lists and village identification*. Agricultural Systems of Papua New Guinea Working Paper Nº 11. Canberra: Human Geography Department, A.N.U.

BRIDGES, L., 1951. *Uttermost Part of the Earth*. London: Hodder & Stoughton.

BROWN, D. E., 1991. *Human Universals*. Boston, Massachusetts: McGraw Hill Inc.

BUROV, G. M., 1967. *Drevnij Sindor*. Moscow : Nauka.

BUROV, G. M., 1998. The Use of Vegetable Materials in the Mesolithic of Northeast Europe. *In*: M. Zvelebil, R. Dennell, L. Domanska, eds. *Harvesting the sea, farming the forest. The Emergence of Neolithic Societies in the Baltic Region*. Archaeological Monographs 10. Sheffield: Sheffield Academic Press.

CAREY, M., 1998. Gender in African beadwork. *In*: L. Sciama, J. B. Eicher, eds. *Beads and Bead Makers, Gender and Material Culture and Meaning*. Oxford: Berg.

CHAPMAN, R., 1868. Available from: http://www.ropecord.com/cordage/biblio/bibliography.

CLAASSEN, C., 1998. *Shells*. Cambridge: Cambridge University Press.

CLARK, D. L., 1976. Mesolithic Europe: the economic basis. *In*: G. de G. Sieveking, I. H. Longworth, K. E. Wilson, eds. *Problems in*

Economic and Social Archaeology. London: Duckworth, 449-481.

CLARK, J. D. G., 1952. *Prehistoric Europe; the economic basis*. London: Methuen.

COLLINS ENGLISH DICTIONARY, 1999. England: Harper Collins.

DECKER, A.-M., 2000. *Nålbinding*. Available from: http://www.geocities.com/sigridkitty/ (Accessed 2007).

DOMINGUEZ-RODRIGO, M., SERRALLONGA, J., JUAN-TRESSERRAS, J., ALCALA, L. and LUQUE, L., 2001. Woodworking activities by early humans: a plant residue analysis on Acheulean stone tools from Peninj (Tanzania). *Journal of Human Evolution*, 40 (4), 289-299.

EMMONS, G. T., 1991. *The Tlingit Indians*. University of Washington Press: Anthropological Papers of the American Museum of Natural History.

GABIOLE, L., 2004. *The Gift of Cowrie*. Available from: http://theearthcenter.com (Accessed June 2005).

GARTH TAYLOR, J., 1974. *Netsilik Eskimo Material Culture. The Roald Amundsen Collection from King William Island*. Oslo: Norwegian Research Council for Science and the Humanities.

GOOD, I., 2001. Archaeological Textiles: A Review of Current Research. *Annual Review of Anthropology*, 30, 209-226.

GRAMSCH, B., 1992. Friesack Mesolithic Wetlands. *In*: B. Coles, ed. *The Wetland Revolution in Prehistory*. The Prehistoric Society. WARP. Occasional Paper 6. Exeter: University of Exeter.

GRØN, O., 1998. Neolithization in Southern Scandinavia. A Mesolithic Perspective. *In*: M. Zvelebil, R. Dennell, L. Domanska, eds. *Harvesting the sea, farming the forest. The Emergence of Neolithic Societies in the Baltic Region*. Sheffield Archaeological Monographs 10. Sheffield: Sheffield Academic Press, 181-191.

HAMPTON, O. W., 1999. *Culture of Stone. Sacred and Profane Uses of Stone among the Dani*. Texas: A&M University Press.

HARDY, K., 2007. Food for thought: Starch in Mesolithic diet. *Mesolithic Miscellany*, 18, 2.

HARDY, K., in press a. Worked bone from Sand. *In*: K. Hardy, C. R. Wickham-Jones, eds.

HARDY, K., in press b. Worked and modified shell from Sand. *In*: K. Hardy, C. R. Wickham-Jones, eds.

HARDY, K., in press c. Prehistoric String Theory. How twisted fibres helped to shape the world. *Antiquity*.

HARDY, K., forthcoming. *Prehistoric carrying capacity. How to get the food home*.

HARDY, K. and SILLITOE, P., 2003. Material

Perspectives: Stone Tool Use and Material Culture in Papua New Guinea. *Available from:* http://intarch.ac.uk/journal/issue14.

HARDY, K. and WICKHAM-JONES, C. R., in press. *Mesolithic and later sites around the Inner Sound, Scotland: the Scotland's First Settlers project 1998-2004.* Available from: www.sair.org.uk.

HARRINGTON, M. R., 1924. The Ozark bluff-dwellers. *American Anthropologist,* 26 (1), 1-21.

HEIDER, K. G., 1970. *The Dugum Dani.* Chicago: Aldine Publishing Company.

HENRY, D. O., HIETALA, H. J., ROSEN, A. M., DEMIDENKO, Y. E., USIK, V. I. and ARMAGAN, T. L., 2004. Human Behavioral Organization in the Middle Paleolithic: Were Neanderthals Different? *American Anthropologist,* 106, 17-31.

HENSHILWOOD, C., D'ERRICO, F., VANHAEREN, M., VAN NIEKERK, K. and JACOBS, Z., 2004. Middle Stone Age Shell Beads from South Africa. *Science,* 304 (5669), 404.

HOWARD, M. C., ed., 2006. *Bark-cloth in Southeast Asia.* Studies in the Material Cultures of Southeast Asia Nº 10. Bangkok: White Lotus Co Ltd.

INDREKO, R., 1967. *Die Mittlere Steinzeit im Estland.* Stockholm: Almqvist snf Wiksells boktr.

JACKSON, J W., 1917. *Shells as evidence of the migrations of early culture.* Manchester and London.

KIMURA, D., 1996. Sex, sexual orientation and sex hormones influence human cognitive function. *Current Opinion in Neurobiology,* 6, 259-263.

KOLLER, J., BAUMER, U. and MANIA, D., 2001. High-tech in the middle Palaeolithic: Neandertal-manfactured pitch identified. *European Journal of Archaeology,* 4 (3), 385-397.

KUHN, S. L., STINER, M. C., REESE, D. S. and GÜLEÇ, E., 2001. Ornaments of the earliest Upper Paleolithic: New insights from the Levant. *Proceedings of the National Academy of Sciences of the United States of America,* 98, 7641-7646.

LEE, R.B., 1979. *The !Kung San: Men, Women and Work in a Foraging Society.* Cambridge: Cambridge University Press.

LEROI-GOURHAN, A. 1982. The archaeology of Lascaux cave. *Scientific American,* 246 (6), 80-88.

LILLIE, M., ZHILIN, M., SHAVCHENKO, S. and TAYLOR, M., 2005. Carpentry dates back to Mesolithic. *Antiquity,* 79 (305). Available from: http://www.antiquity.ac.uk/projgall/lillie/index.html.

MACKENZIE, M., 1991. *Androgynous Objects: String Bags and Gender in Central New Guinea.* Chur: Harwood Academic Publishers.

MAIR, L., 1969. *Witchcraft.* London: Weidenfield and Nicholson.

MALINOWSKI, B., 1922. *Argonauts of the Western Pacific.* London: Routledge & Sons.

MAZZA, P. P. A., MARTINI, F., SALA, B., MAGI, M., COLOMBINI, M. P., GIACHI, G., LANDUCCI, F., LEMORINI, C., MODUGNO, F. and RIBECHINI, E., 2006. A new Palaeolithic discovery: tar-hafted stone tools in a European Mid-Pleistocene bone-bearing bed. *Journal of Archaeological Science,* 33 (9), 1310-1318.

MELLARS, P., 1987. *Excavations on Oronsay. Prehistoric Human Ecology on a small island.* Edinburgh: Edinburgh University Press.

MERTENS, E.-M., 2000. Linde, Ulme, Hasel, Zur Verwendung von Pflanzen für Jagd- und Fischfanggeräte im Mesolithikum Dänemarks und Schleswig-Holsteins. *Prehistorische Zeitschrift,* 75, 1-55.

MITHEN, S., 2000. *Hunter-gatherer landscape archaeology. The Southern Hebrides Mesolithic Project 1988–98. Vols. 1 and 2.* Cambridge: McDonald Institute Monograph Series.

MORDANT, D. and MORDANT, C., 1992. Noyen-sur-Seine: A Mesolithic Waterside Settlement. *In:* B. Coles, ed. *The Wetland Revolution in Prehistory.* The Prehistoric Society. WARP. Occasional Paper 6. Exeter: University of Exeter, 56-64.

MYKING, T., HERTZBERG, A. and SKRØPPA, T., 2005. History, manufacture and properties of lime bast cordage in northern Europe. *Forestry,* 78 (1), 65-71.

NADEL, D., DANIN, A., WERKER, E., SCHICK, T., KISLEV, M. E. and STEWART, K., 1994. 19,000-Year-Old Twisted Fibers From Ohalo II. *Current Anthropology,* 35 (4), 451-458.

Nelson, E. W., 1983. *The Eskimo about Bering Strait.* Washington: Smithsonian Institution Press.

OWEN, L. R., 1993. Material worked by hunter and gatherer groups of northern North America: implications for use-wear analysis. *In:* P. Anderson, S. Beyries, M. Otte and H. Plisson, eds. *Traces et fonction : les gestes retrouvés. Actes du colloque international de Liège, 8-10 Décembre 1990.* Liège: ERAUL 50, 3-12.

PAIJMANS, K., 1976. *New Guinea Vegetation.* Canberra: Australia National University Press.

PARKS, R. L. and BARRETT, J. H., in press. The Zooarchaeology of Sand. *In:* K. Hardy, C.R. Wickham-Jones, eds.

RIMANTIENÉ, R. 2005. *Akmens Amžiaus žvejai prie Pajūrio Lagūnos.* Vilnius: Lietuvos Nacionalinis Muziejus, 507-526.

SCIAMA, L. and EICHER, J. B., 1998. *Beads and bead makers: gender, material culture, and meaning.* Oxford, New York: Berg.

SILLITOE, P., 1988. *Made in Niugini.* London: British Museum Publications.

SILLITOE, P. and HARDY, K., 2003. Living Lithics: ethnoarchaeology in Highland Papua New Guinea. *Antiquity*, 77 (297), 555-566.

SKAARUP, J., 1982. Sites 7, Skoldnaes & 8 Derjø. Recent excavations and discoveries. *Journal of Danish Archaeology*, 1.

SKAARUP, J. and GRØN, O., 2004. *Møllegabet II. A submerged Mesolithic settlement in southern Denmark*. British Archaeological Report International Series 1328 & Langelands Museum. Oxford: Archeopress.

SOFFER, O., 2004. Recovering Perishable Technologies through Use Wear on Tools: Preliminary Evidence for Upper Palaeolithic Weaving and Net Making. *Current Anthropology*, 45 (3), 407-413.

SOFFER, O, ADOVASIO, J. M. and Hyland, D. C., 2000. The « Venus » Figurines. Textiles, Baskery. Gender and Status in the Upper Palaeolithic. *Current Anthropology,* 41 (4), 511-537.

SOFFER, O, ADOVASIO, J. M. and Hyland, D. C., 2001. Perishable Technologies and Invisible People: Nets, Baskets and « Venus » wear ca. 26,000 BP. *In*: B. Purdy, ed. *Enduring Records: The Environmental and Cultural Heritage*. Oxford: Oxbow Books, 233-245.

STEEL, T., 1994. *The Life and Death of St Kilda*. London: Harper Collins.

THOMSON, D. F., 1936. Notes on Some Bone and Stone Implements from North Queensland. *The Journal of the Royal Anthropological Institute of Great Britain and Northern Ireland*, 66, 71-74.

VANHAEREN, M., D'ERRICO, F., STRINGER, C., JAMES, S. L., TODD, J. A. and MIENIS, H. K., 2006. Middle Paleolithic Shell Beads in Israel and Algeria. *Science,* 23 (312), 1785-1788.

VOGT, E., 1937. *Geflechte und Gewebe der Steinzeit*. Basel: Birkhauser.

WARNER, C. and BEDNARIK, R., 1996. Pleistocene Knotting. *In*: J. C. Turner and P. van de Griend, eds. *History and Science of Knots*. Singapore: World Scientific, 3-18.

ZVELEBIL, M., 1994. Plant use in the Mesolithic and its role in the transition to farming. *Proceedings of the Prehistoric Society*, 60, 35-74.

PRÉHISTOIRE DU TRAVAIL DES PLANTES DANS LE NORD DE LA BELGIQUE. LE CAS DU MÉSOLITHIQUE ANCIEN ET DU NÉOLITHIQUE FINAL EN FLANDRE

Valérie BEUGNIER

Résumé : Comme en attestent les résultats obtenus lors de l'étude fonctionnelle de trois sites du nord de la Belgique, le travail des plantes a constitué pour certains groupes européens une activité majeure. Dans tous ces sites, le raclage des matières végétales est, en effet, la fonction dominante de l'outillage en silex. À Verrebroek et Doel, campements de chasse datés du Mésolithique ancien, on a ainsi les premiers témoins d'une exploitation massive des plantes à l'aide d'outils en silex. À Waardamme, site Néolithique final du Deûle/Escaut, en revanche, on aborde la question des microdenticulés et de leur énigmatique fonction. Cette recherche a, par ailleurs, permis de mettre en évidence les problèmes spécifiques rencontrés lors de l'interprétation en termes d'activités des traces d'utilisation liées au travail des plantes. C'est aussi l'occasion de montrer en quoi une telle recherche, fondée en partie sur l'expérimentation, ne peut être menée à bien, compte tenu du niveau de savoir-faire et de connaissances techniques requis, que par la mise en place de collaborations effectives entre différents spécialistes, tels que tracéologues, ethnologues, ethno-historiens, botanistes et artisans.

Mots-clés : trace d'utilisation, raclage, plante, microdenticulés, Mésolithique ancien, Néolithique final.

Abstract: Use-wear analysis of flint tools excavated at three sites in northern Belgium has demonstrated that the role of plant processing was very important in prehistoric societies of western Europe. On all three sites plant scraping is the most important activity performed by means of lithic tools. Verrebroek and Doel temporary camps of small groups of hunter-gatherers dated to the early Mesolithic (Pre-boreal – Boreal) have yielded the oldest indirect evidence of the use of flint tools for plant processing activities. At the final Neolithic settlement of Waardamme, belonging to the so-called Deûle/Escaut group of northern France, microdenticulates or serrated-edge tools were mainly used for the same purpose. In this paper the problems encountered during the behavioural interpretation of plants wear traces will be discussed. Due to the high level of technical skills and know-how involved, these problems can only be addressed through a multi-disciplinary approach, with use-wear analysts, ethnologists, botanists and craftsmen working together.

Keywords: usewear, scraping, plant, microdenticulates, early Mesolithic, final Neolithic.

INTRODUCTION

L'importance du travail des plantes dans les sociétés néolithiques est maintenant bien établie. En Europe, c'est notamment à partir des découvertes réalisées en milieu lacustre que la richesse et la diversité de ce domaine technique ont pu être déterminées. Ces sites d'habitat, datés pour la plupart de la fin du Néolithique, ont, en effet, livré des milliers d'objets et de produits élaborés à partir de végétaux : aliments, fourrages, éléments de paroi et de toiture, vêtements, chaussures, paniers, filets, cordes, outils, armes, etc. (Pétrequin 1984, Coles and Lawson 1987, Pétrequin et Pétrequin 1988, Coles and Coles 1989). Dans le même temps, les études fonctionnelles d'outillages en silex confirmaient l'importance des activités de récolte et de traitement des matières végétales dans les sociétés d'agriculteurs-éleveurs (par exemple, Anderson 1992, Juel Jensen 1994, Gassin 1996, Astruc 2002).

À ce jour, plus personne ne s'étonne ainsi de la forte présence des plantes au sein des économies de production. Concernant le Mésolithique, en revanche, il est plus difficile de se faire une idée précise de la place occupée par ce domaine technique. En Europe septentrionale, des découvertes réalisées en contexte humide montrent pourtant, que dès cette époque, il existait un artisanat des plantes à l'origine d'une production variée. Le site de Friesack, en Allemagne de l'Est, daté du Mésolithique ancien, a fourni les plus anciens témoins directs du travail des plantes, sous la forme de fragments de flotteurs en écorce de bouleau mais aussi de cordelettes, ficelles et filets, fabriqués à partir de filasse d'écorce de saule notamment (Gramsch 1987, Körber-Grohne 1995). D'autres sites plus récents ont aussi montré toute la richesse de cet artisanat. Par exemple, au Danemark, dans les sites de Skjoldnaes et Tybrind Vig, étaient conservés des fragments de vanneries, de filets, de textiles, de fils et de cordes, la plupart élaborés à partir de baguettes et de liber d'écorce de saule (Skaarup 1983, Andersen 1987). À Bergschenhoek et à Schipluiden, aux Pays-Bas, ont aussi été retrouvés trois grandes nasses à poisson tressées avec de fines tiges de cornouiller (Louwe Koojimans 1987) et des fragments de tissus et de vanneries (Kooistra 2006). Plus encore, une étude tracéologique récente, portant sur ces séries du nord de la Belgique, a permis d'établir que le travail des matières végétales souples a pu constituer, dès le début du Mésolithique, la fonction dominante des outillages en silex (Beugnier et Crombé 2005).

Dans cette région et plus largement dans toute l'Europe du Nord, nous avons ainsi l'opportunité de suivre, notamment à partir de l'étude des industries lithiques taillées, les caractéristiques et les évolutions d'un artisanat sur plusieurs millénaires, des premiers temps de l'occupation mésolithique à la fin du Néolithique. La reconstitution des artisanats et des techniques préhistoriques liés au travail des plantes reste néanmoins un problème majeur auquel se heurtent toutes les analyses tracéologiques menées jusqu'à présent. Dans la plupart des cas, il n'est, en effet, pas possible d'établir avec précision quelle activité a été menée ni même quel matériau a été travaillé. À cela, différentes raisons liées aux limites mêmes de la méthode mais aussi aux spécificités du monde végétal : grande diversité des plantes potentiellement utilisables, grande variété des techniques ayant pu être mises en œuvre pour des finalités également très différentes mais aussi, matériaux périssables rarement conservés dans les sites et domaine d'activité relativement discret, peu valorisé et peu spectaculaire et donc peu connu qui n'a suscité qu'un intérêt limité aussi bien chez les ethnologues que chez les archéologues.

Afin d'illustrer ce propos, deux études de cas seront maintenant présentées. Celles-ci concernent des assemblages du nord de la Belgique (fig. 1), composés d'un nombre conséquent d'outils marqués par des usures d'origine végétale, posant d'importants problèmes d'interprétation tant pour la reconnaissance de la matière travaillée que pour celle de l'activité réalisée. Dans le premier exemple, les séries étudiées proviennent de campements de chasseurs-cueilleurs mésolithiques et ont fourni les plus anciens témoins d'une exploitation massive des plantes à l'aide d'outils en silex. Le second exemple appartient, quant à lui, au Néolithique final et traite de l'énigmatique question des microdenticulés.

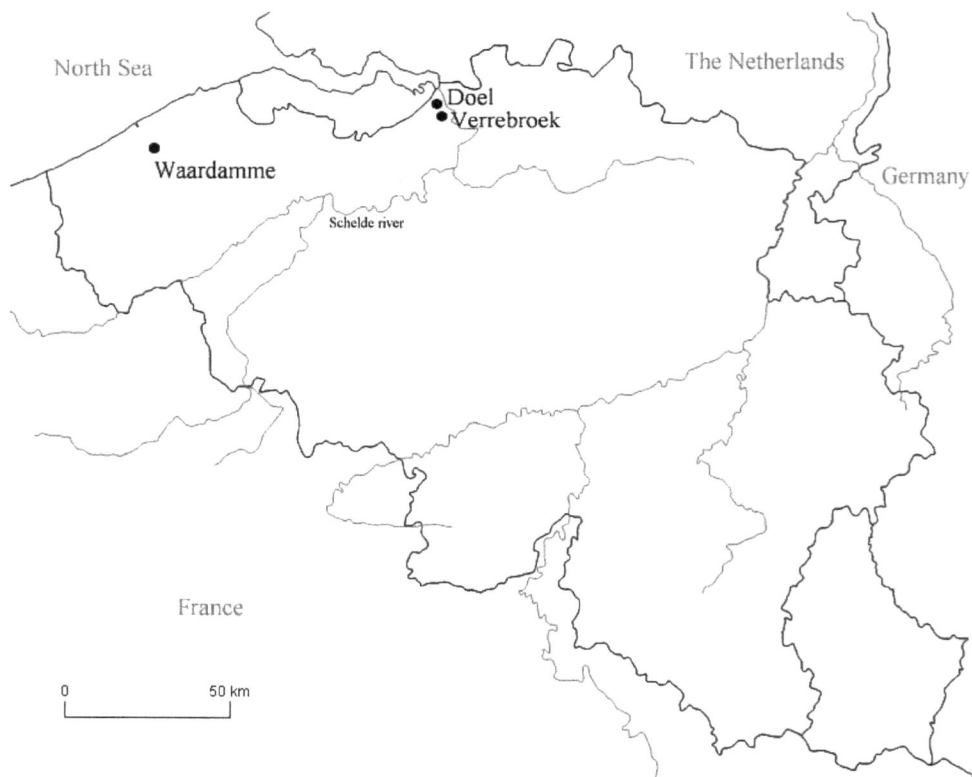

Figure 1. Implantation des sites de Verrebroek et Doel du Mésolithique ancien et de Wardamme du Néolithique final, dans la région des sols sableux du nord de la Belgique.

LE TRAITEMENT DES MATIÈRES VÉGÉTALES AU MÉSOLITHIQUE ANCIEN : LES PREMIERS INDICES D'UNE UTILISATION MASSIVE DU SILEX

Les sites de Verrebroek et Doel, distants d'environ cinq kilomètres l'un de l'autre, sont localisés en Flandre orientale dans la région des sols sableux, à proximité des rives de l'Escaut. Implantés en zone humide sur d'anciennes dunes, ces gisements ont bénéficié de conditions de conservation favorables, ayant été scellés après abandon par des niveaux de tourbes et des dépôts alluviaux. Ils ont ainsi fourni un matériel relativement abondant et varié, retrouvé en place. Menacés par les travaux d'agrandissement du port d'Anvers, ces sites ont fait l'objet entre 1992 et 2003, d'un programme de fouilles de sauvetage placé sous la direction de Ph. Crombé (1998, 2005, Bats *et al.* 2003).

Le site de Verreborek a été fréquenté à plusieurs reprises durant le Mésolithique ancien entre 8740 et 7560 cal. BC (avec 95 % de probabilité), soit de la seconde moitié du Pré-Boréal à la première moitié du Boréal (Van Strydonck *et al.* 2001). Dans ce site, les différentes campagnes effectuées ont permis de couvrir une surface de 6000 m² qui représente environ 20 % de la surface totale du site, estimée à plus de 3 hectares. Sur toute cette zone de fouille, le matériel est réparti en 55 concentrations ou unités lithiques comprenant chacune entre un et cinq foyers, des silex et de rares matières organiques présentes uniquement sous forme carbonisée et constituées de coquilles de noisettes, de charbons de bois et d'os (Crombé *et al.* 2003). L'industrie lithique essentiellement aménagée sur des silex locaux de mauvaise qualité est largement dominée par les microlithes qui, comme une étude tracéologique a permis de le démontrer, ont servi de façon quasi-exclusive à armer des projectiles, soit comme barbelures, soit comme pointes (Crombé *et al.* 2001).

D'autres types d'outils ont également été retrouvés mais ils sont beaucoup moins nombreux et ne sont représentés qu'au sein de certains ensembles. Il s'agit de grattoirs, de burins, de perçoirs, de lames et d'éclats diversement retouchés. On note également que de très grandes quantités de déchets liés, entre autres, au débitage et au façonnage des microlithes ont été récoltées dans certaines concentrations, témoignant de l'existence d'activités de taille parfois importantes autour des foyers.

Dans le détail, une certaine variabilité a pu être mise en évidence dans les dimensions, la densité et le contenu de ces concentrations. On identifie ainsi des ensembles de petite taille (< 30 m²) et d'autres plus larges (> 46 m²), des ensembles à faible densité de matériel, comprenant entre 1500 et 3500 silex et d'autres à forte densité de matériel, comprenant plus de 15000 silex. Sur le plan typologique également, la représentation des différents types d'outils peut varier d'une concentration à l'autre. Dans certains assemblages, les microlithes et les grattoirs représentent entre 25 et 50 % du matériel retouché. Dans d'autres, les grattoirs ne forment plus qu'un pourcentage très faible alors que les armatures sont surreprésentées. Enfin, de très petites concentrations ne sont constituées que de quelques pièces retouchées. Ces différences dépendraient, si l'on en croit les résultats des études tracéologiques menées, des durées d'occupation et/ou de la taille des groupes ayant occupé ces différents espaces (Beugnier et Crombé 2005). Verrebroek correspondrait ainsi à une succession d'occupations saisonnières, de durée variable et/ou concernant des groupes de taille différente. Ces implantations sont, de toute évidence, « spécialisées », autour au moins de quatre activités clairement identifiées, la chasse, la fabrication de l'armement en pierre, le travail des plantes et dans une moindre mesure le traitement des peaux (*cf. infra*).

À quelques kilomètres de là, le site de Doel, s'étendant sur une surface beaucoup plus réduite, correspond plus probablement aux vestiges d'une seule occupation. Il a livré une petite concentration lithique que trois dates ¹⁴C, réalisées sur des coquilles de noisette brûlées, permettent de raccorder au Mésolithique ancien et à la première moitié du Boréal.

L'étude fonctionnelle a porté sur un échantillon de 461 pièces prélevées à Doel et au sein de huit concentrations du site de Verrebroek (Beugnier et Crombé 2005, Beugnier 2006a, b). Cet échantillon comprend au total une sélection de 191 outils et de 270 lames, lamelles et éclats bruts (fig. 2). Comme en attestent les résultats obtenus, différentes activités ont été menées, dans ces sites, en plus de la chasse (fig. 3). Parmi celles-ci, le travail des plantes non ligneuses domine largement et cela dans toutes les concentrations analysées, à deux exception près où cette activité n'est soit pas représentée (C35), soit devancée par le traitement des peaux (C23).

Une telle importance du travail des plantes a constitué un résultat inattendu. Verrebroek et Doel sont, en effet, pour toute l'Europe du Nord, les

	Concentrations	Nombre de silex	Nombre d'outils et de microlithes	Nombre de pièces analysées	Nombre d'outils analysés
Concentrations A (*25 à 50 % de grattoirs et 20 à 50 % microlithes*)	C23	2928	58	40	19
	C24	1450	31	25	5
	C70	26 223	247	95	50
	C22	42 993	199	124	59
Concentrations B (*50 à 90 % de microlithes*)	C59	3407	45	32	5
	Doel C3	14 500	99	75	25
Concentrations C (*outils rares*)	C33	674	11	11	5
	C35	883	3	15	1
	C34	1194	10	22	2
Outils hors concentrations	-	-	-	20	20

Figure 2. Échantillons observés au microscope, prélevés à Doel et dans huit concentrations du site de Verrebroek.

Domaines d'activités représentés	C23	C24	C70	C22	C59	Doel C3	C33	C35	C34	Total
Traitement des plantes non ligneuses	1	6	25	27	9	9			11	88
Travail du bois	1		1	2		2				6
Traitement des peaux	3	3	22	12	1	4		2		47
Travail de l'os				4						4
Bricuets	1		2		1					4
Divers	1		3							4
Matières indéterminées	5	5	14	10	2	17		1	2	56
Total	12	14	67	55	13	32		3	13	209

Figure 3. Spectre fonctionnel de l'outillage en silex provenant du site de Doel
et de huit concentrations du site de Verrebroek (décomptes en nombre de zones utilisées identifiées).

sites les plus anciens à témoigner d'une utilisation massive du silex pour le traitement des matières végétales. Jusqu'à présent, le travail des végétaux n'était attesté que ponctuellement en contexte Mésolithique ancien, comme par exemple à Zutphen aux Pays-Bas (Groenenwoudt *et al.* 2001) ou dans d'autres régions comme dans les sites sauveterriens et montclusiens du Sud de la France (Philibert 2002). Après cela, il faut attendre le Mésolithique final et récent pour retrouver des assemblages ayant joué un rôle non négligeable dans la préparation de diverses plantes siliceuses. Par exemple, on pense aux sites néerlandais de Swifterbant (Bienenfeld 1985), Almere (Peeters *et al.* 2001) et Hardinxveld-

Giessendam Polderweg et De Bruin (Van Gijn *et al.* 2001a, b) ainsi qu'aux sites scandinaves d'Ageröd V (Larsson 1983) et Vaenget Nord (Juel Jensen and Brinch Petersen 1985). Verrebroek et Doel représentent bien ainsi les premiers sites attribués à la transition Préboréal/Boréal où le travail des plantes constitue une des fonctions principales de l'outillage en silex.

Dans ces deux sites, parmi l'ensemble des outils en contact avec une matière végétale souple, seuls dix supports ont servi selon un mouvement longitudinal, à couper, fendre, inciser et percer (fig. 4). Les 88 autres ont fonctionné selon un *mouvement transversal oblique*.

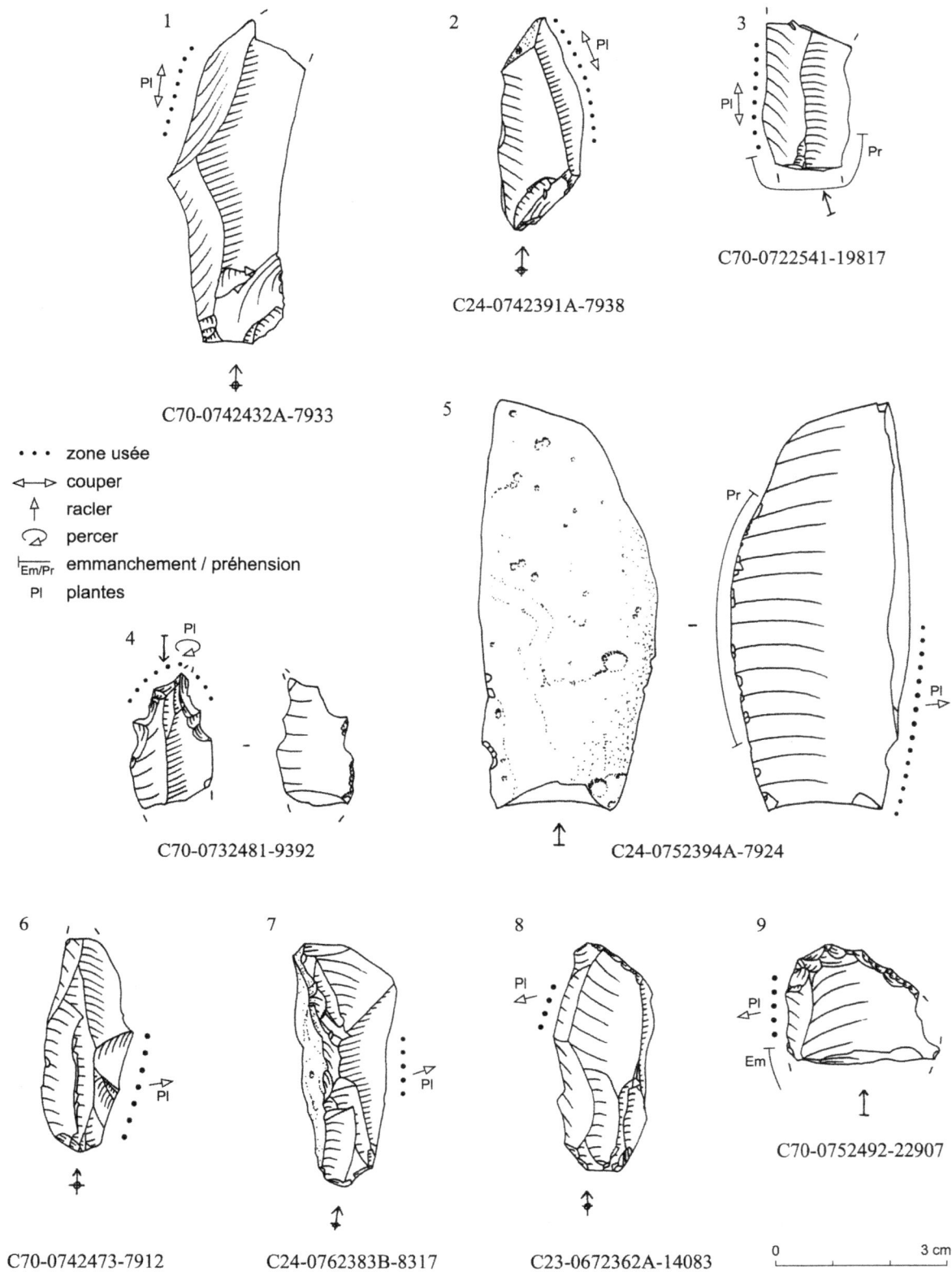

Légende :
- • • • zone usée
- ◁—▷ couper
- ↑ racler
- ⌒ percer
- Em/Pr emmanchement / préhension
- Pl plantes

1 — C70-0742432A-7933
2 — C24-0742391A-7938
3 — C70-0722541-19817
4 — C70-0732481-9392
5 — C24-0752394A-7924
6 — C70-0742473-7912
7 — C24-0762383B-8317
8 — C23-0672362A-14083
9 — C70-0752492-22907

0   3 cm

Figure 4. Exemple d'outils en silex du site de Verrebroek présentant diverses usures dues au contact avec des plantes non ligneuses.

Les outils utilisés longitudinalement ont travaillé des plantes d'origine indéterminée, dans le cadre probable d'opérations de traitement ou de transformation de la matière plutôt que pour la collecte des tiges. Les polis observés, sur ces exemplaires, sont peu intenses et peu caractéristiques et le matériel sélectionné est composé de lames, de lamelles et d'éclats bruts, de gabarit réduit. Ainsi, à Verrebroek et Doel, il n'existerait pas d'instruments en silex réservés à la récolte des plantes pourtant abondamment exploitées dans ces sites.

Les outils de raclage se répartissent en deux ensembles. Le premier groupe est formé d'une quinzaine d'éléments présentant un poli brillant, uni, fluide et légèrement cannelé, assez proche des polis dus au raclage expérimental de tiges de roseaux (*Phragmites*), de joncs (*Juncus*) et de massettes (*Typha*) (fig. 5). Ces quelques pièces ont ainsi pu servir dans le cadre de chaînes opératoires de fabrication ou de restauration d'objets ou d'armes. À l'image des expérimentations menées, on peut, par exemple, envisager une utilisation de ces instruments pour fabriquer des fûts de flèche à partir de tiges de roseaux, régularisées par raclage au silex.

L'autre groupe largement dominant (n = 63) est marqué par des usures indéterminées. Sur ces outils, la face en contact avec la matière est caractérisée par un poli généralement étendu et souvent visible à l'œil nu, *brillant, uni, plat et très fortement strié* (fig. 6).

Les stries sont longues et profondes et se développent selon un angle de plus ou moins 45° par rapport au bord actif. Le poli présente parfois une certaine variabilité, passant le long d'un même bord d'outil, d'une coalescence plate striée à une coalescence non striée, plus fluide, parfois légèrement cannelée, assez proche des polis de roseaux. Sur la face opposée, on observe les mêmes usures ou, à l'inverse, un poli marginal, uni, plat/concave. L'outillage utilisé est géné-

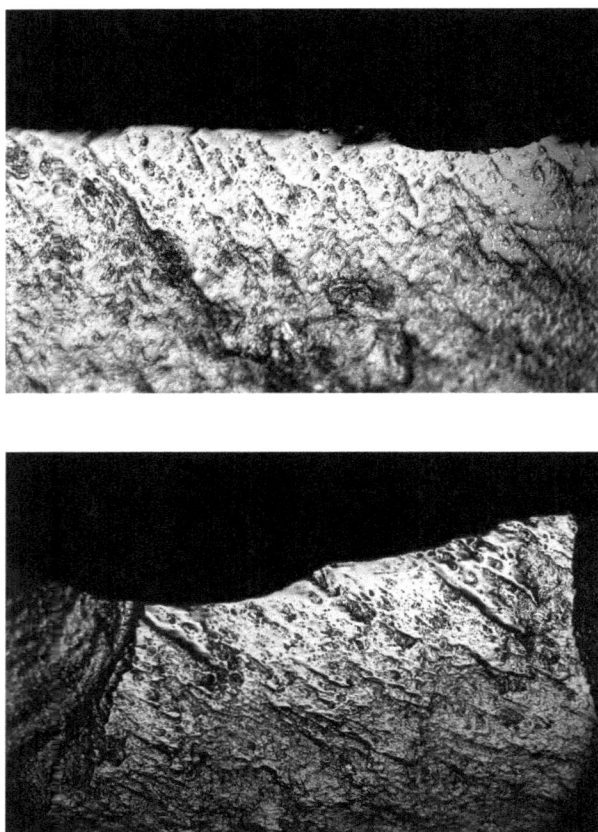

Figure 5. Exemples d'usures archéologique et expérimentale, liées au travail des plantes souples à forte teneur en silice. En haut, éclat brut (O74 Z49 1A 8031) de la concentration C70 du site de Verrebroek (X200). En bas, pièce expérimentale utilisée pour fendre des massettes à l'état vert (*Typha*) (X100).

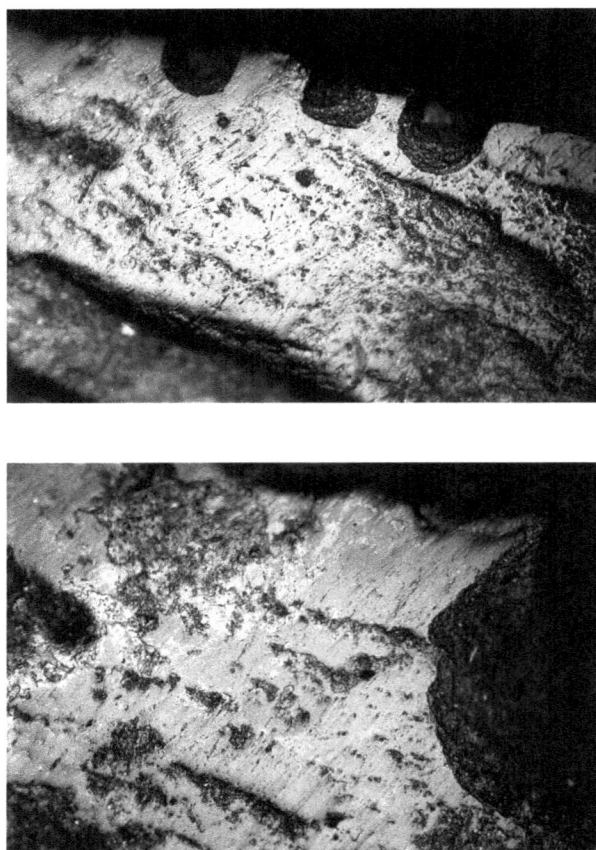

Figure 6. Grattoir (O75 Z492 22907) de la concentration C70 du site de Verrebroek. Détails d'un poli de raclage dû au contact avec une plante indéterminée à forte teneur en silice (X200).

ralement léger. Il est composé essentiellement de lames, de lamelles et d'éclats bruts à bords fins.

Des stigmates similaires ont été observés lors de l'étude de deux autres séries d'âge plus récent (fig. 7). Celles-ci proviennent du site hollandais de Hardinxveld-Giessendam Polderweg (van Gijn *et al.* 2001a) ainsi que du site belge de Doel, secteur B (Beugnier 2007), datant tous deux du Mésolithique final (culture de Swifterbant). On notera que ces deux sites sont aussi des campements de chasseurs-cueilleurs de tradition mésolithique, géographiquement proches, situés en milieu humide et appartenant à la région des sols sableux s'étendant du nord de la Belgique aux Pays-Bas.

nasses, les paniers, etc. Dans le cas du site de Hardinxveld-Giessendam Polderweg, l'idée selon laquelle les usures végétales observées pourraient être dues à la préparation de tubercules aquatiques comestibles a également été considérée. Dans quelques sites hollandais, des traces de tubéreuses consommables ont, en effet, été retrouvées dans des foyers. Il s'agit de racines de massette (*Typha*), scirpe (*Scirpus*), fougère (*Dryopteris filix-mas*), blette (*Beta vulgaris* ssp. *maritima*) et prêle (*Equisetum*) (Perry 1999). En Belgique, aucun foyer mésolithique n'a, en revanche, livré de reste carbonisé de plante comestible, à l'exception des noisettes (*Corylus avellana*) et du chénopode (*Chenopodium album*) (Meersschaert 2005). Enfin, suite aux récents travaux menés par K. Hardy (ce

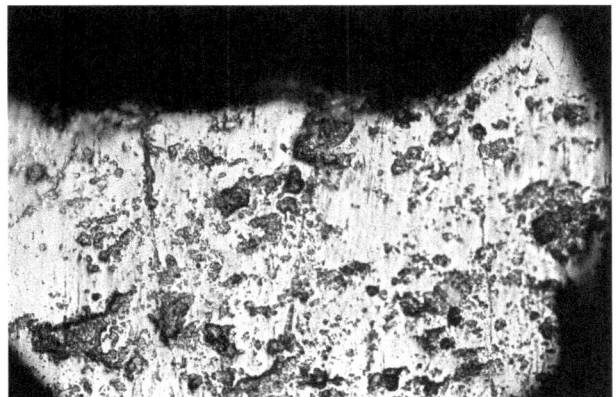

Figure 7. Hardinxveld-Giessendam Polderweg, pièce n° 13354. Détails d'un poli de raclage dû au contact avec une plante à forte teneur en silice, similaire aux polis observés sur le matériel des sites du Mésolithique ancien de Verrebroek et Doel (X100, X200).

En dehors de ces contextes, aucune autre référence à ces usures particulières n'a pu être trouvée. Ainsi, à la différence de certains polis végétaux de plus large répartition (*cf. infra*, le « poli des micro-denticulés » par exemple), les traces en question pourraient témoigner de l'existence d'une techni-que de travail des plantes peu répandue, strictement limitée aux groupes mésolithiques de la région des deltas du Rhin, de la Meuse et de l'Escaut. Compte tenu de leurs caractéristiques, ces traces sont supposées résulter du contact avec une matière végétale tendre à forte teneur en silice, travaillée selon un mouvement de raclage visant à enlever des copeaux de matière ou selon un mouvement de fendage visant à séparer la matière.

Sur cette base et en tenant compte du contexte archéologique, différentes hypothèses fonction-nelles peuvent être proposées telles que la fabrication d'armes ou d'objets comme les hampes de flèche, les

volume) et L. Hurcombe (ce volume), l'utilisation de ces outils dans le cadre de chaînes opératoires d'extraction des fibres pour la production de fil et de ficelle est apparue comme une nouvelle hypothèse extrêmement séduisante qu'il conviendrait de tester le plus rapidement possible.

On constatera que les analyses palynologiques réalisées dans les environs immédiats du site de Doel (Deforce *et al.* 2005) ont montré qu'un grand nombre de plantes étaient disponibles durant l'Atlantique et le début du Subboréal (Mésolithique récent/final). Ce site est, en effet, implanté au cœur d'une zone humide et à proximité de forêts, soit un envi-ronnement où coexiste une grande variété d'espèces forestières et aquatiques : noisetier (*Corylus*), tilleul (*Tilia*), saule (*Salix*), bouleau (*Betula*), roseau (*Phragmites*), laîche ou carex (Cyperaceae), prêle (*Equisetum*), etc. Les spectres polliniques montrent, en revanche, que les plantes telles que les nénuphars (*Nuphar*) et les lotus *(Nymphaea)*, croissant en eau plus profonde et dont les racines auraient pu être consommées, font défaut.

Sur le plan expérimental, un certain nombre de ces matériaux ont déjà été testés par moi-même ou d'autres chercheurs[1], soit de façon mécanique, soit dans le cadre de chaînes opératoires en rapport avec la fabrication de hampes de flèche ou la réalisation de sparteries. Ont ainsi été travaillés par raclage au silex : des roseaux (*Phragmites australis*), des joncs (*Juncus*), des massettes (*Typha*), des graminées sauvages, des laîches (Cyperaceae), des tiges de noisetier (*Corylus avellane*) et de viorne (*Viburnum lantana*), des ronces (*Rubus*), des orties (*Urtica dioica*), des fougères (*Dryopteris*) et de la filasse d'écorce de tilleul (*Tilia*). L'observation de ces pièces au microscope ne m'a pas permis de retrouver des usures semblables à celles relevées sur le matériel archéologique. Il est, de cette façon, impossible de préciser quel matériau a été travaillé et encore moins quelle chaîne opératoire a été mise en œuvre.

Cet échec peut résulter du nombre trop faible d'expériences menées de façon ciblée et systématique, en rapport avec les hypothèses proposées. Comme cela a déjà été dit, le travail des plantes recouvre des pratiques d'une très grande diversité dont ne rendent pas compte les référentiels expérimentaux actuels. Pour des raisons de temps notamment, on comprendra facilement que chaque expérimentateur n'a pu tester qu'un nombre réduit des multiples possibilités.

Mais, l'inadéquation entre usure archéologique et usure expérimentale peut aussi découler d'une série de facteurs indépendants de l'hypothèse testée. En effet, pour le végétal bien plus que pour les autres matériaux (os, peau, etc.), les caractéristiques des stigmates d'utilisation vont varier de façon conséquente en fonction de l'état de la matière (espèce travaillée mais aussi partie de la plante mise en œuvre, degré de maturité, taux d'humidité, date de récolte, nature du sol, etc.) et des condition d'utilisation des outils (position de l'outil, durée et intensité du travail, etc.)

(Anderson-Gerfaud 1981, Plisson 1985, Vaughan 1985, Grace 1989, van Gijn 1989, Anderson 1992, González Urquijo, Ibáñez Estévez 1994, Clemente Conte 1995, Ibáñez Estévez, González Urquijo 1996, Astruc 2002). Or, ces variables dépendent de façon étroite du niveau de connaissance et du degré d'habileté des expérimentateurs. Ainsi, on peut facilement supposer qu'au cours des tests réalisés, bon nombre d'entre nous peu expérimentés n'ont pas travaillé les bonnes plantes, au bon moment, avec les bons gestes. Comme en témoigne l'article de L. Hurcombe (ce volume), l'écart existant entre contexte expérimental et situation archéologique peut toutefois être réduit par la mise en place de programmes expérimentaux ciblés autour d'une problématique spécifique, permettant de tester différentes hypothèses de façon systématique, tout en acquérant un certain savoir-faire. Le développement de collaborations réunissant différents spécialistes autour de problématiques communes est tout aussi fondamental, particulièrement dans le cas d'un domaine technique pour lequel, à quelques exceptions près, on ne retrouve, dans les sites, aucun témoin direct conservé. Dans ce contexte, artisans, ethnologues et ethno-historiens vont représenter, pour l'élaboration des hypothèses fonctionnelles, une source majeure d'informations.

C'est précisément dans cette perspective de confrontation des données issues de domaines de recherche complémentaires qu'il a, par exemple, été permis de renouveler notre perception de l'outillage utilisé à Verrebroek et Doel. Jusqu'alors, les industries lithiques retrouvées dans des sites dits « spécialisés », correspondant à des campements de courte durée où la chasse est une activité importante, sont d'abord interprétées sous l'angle de cette spécialisation. Ainsi, elles sont de façon préférentielle supposées réservées à la chasse, la fabrication et la restauration de l'armement, la préparation et le traitement des carcasses et des peaux. Or, l'étude de K. Hardy (ce volume) portant sur un réexamen du Méso-lithique européen sous l'éclairage de données rassemblées en Papouasie-Nouvelle-Guinée nous a permis d'envisager une toute autre hypothèse, qui reste bien évidemment à valider sur le plan expérimental. On rappellera ici que cette hypothèse est relative aux chaînes opératoires d'extraction et de production des fibres pour la fabrication de fils, ficelles et cordes.

Dans tous les cas, comme en témoignent les résultats tracéologiques obtenus à Verrebroek et Doel, il s'agit d'une activité de traitement des plantes significative du point de vue économique

---

[1.] Je souhaite remercier ici Pierre Pétrequin qui a réalisé certaines des expérimentations citées, sur la base des découvertes d'objets en matières et fibres végétales réalisées dans les sites lacustres de Chalain et Clairvaux (Jura, France) (Pétrequin 1986, 1989, 1997). Je tiens également à remercier Patricia Anderson, Sylvie Beyries, Bernard Gassin et Annelou van Gijn qui ont mis, dans le cadre de différents projets, leurs séries expérimentales à ma disposition.

et technique qui a été menée de façon répétée au cours de chaque occupation, soit tout au long de l'histoire des sites étudiés. Par ailleurs, les procédés mis en œuvre dans le cadre de cette activité sont, semble-t-il, strictement liés aux cultures mésolithiques du nord de l'Europe et tout particulièrement à celles de la zone des sols sableux du nord de la Belgique et des Pays-Bas.

LE SITE NÉOLITHIQUE FINAL DE WAARDAMME : L'ÉNIGMATIQUE QUESTION DES MICRODENTICULÉS

Comme Verrebroek et Doel, le site de Waardamme, situé en Flandre occidentale, appartient à la région des sols sableux du nord de la Belgique (Demeyere *et al.* 2006) (fig. 1). Il correspond aux vestiges relativement bien conservés d'un site d'habitat du Néolithique final, daté de la première moitié du 3e millénaire av. J.-C. Il est rattaché au groupe culturel régional du Deûle/Escaut dont l'extension se limitait, jusque-là, au nord de la France et au Hainaut occidental belge (Blanchet 1984, Piningre 1985, Martial *et al.* 2004). Waardamme représente ainsi l'expansion maximale de ce groupe en direction des zones septentrionales. Dans ce site, ont été mis au jour les vestiges d'un bâtiment, formé d'une tranchée de fondation dans laquelle étaient implantés des poteaux soutenant les parois extérieures de la maison, faites de clayonnage et de torchis (fig. 8). À l'intérieur du bâtiment, des trous de poteaux et/ou des fosses ont également été mis en évidence. Le sol d'habitat n'est malheureusement pas conservé. Le matériel recueilli était réparti dans la

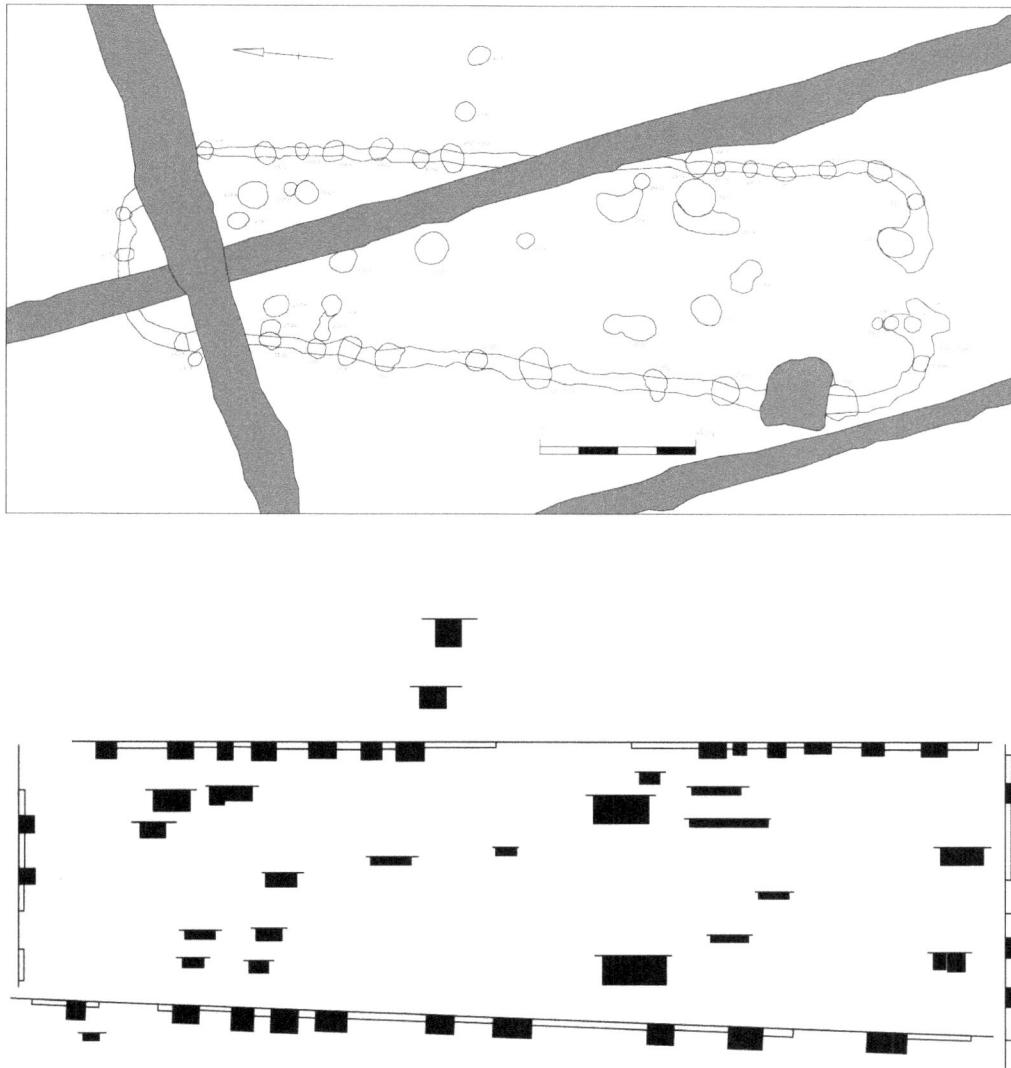

Figure 8. Plan détaillé des structures d'habitat du Néolithique final découvertes lors des fouilles : tranchées de fondation, trous de poteaux et fosses.

tranchée de fondation et les trous de poteau internes. Il comprend un groupe de tessons mais aussi un peson et quatorze fusaïoles liés au traitement des fibres textiles. L'industrie lithique est formée de fragments de hache polie et de grès et d'une industrie en silex taillée, comprenant moins de 300 éléments parmi lesquels on décompte environ 50 outils dont deux pièces en silex du Grand-Pressigny. Comme dans tous les assemblages attribués au Deûle/Escaut, cette industrie se distingue sur le plan typologique par une importante série de microdenticulés représentant presque 50% du matériel retouché.

Le microdenticulé est un outil maintenant assez connu, caractérisé par une série de retouches lui donnant l'aspect d'une micro-scie et la présence d'un poli intense dû au *raclage d'une matière végétale siliceuse indéterminée*. Sur les vingt-six microdenticulés retrouvés à Waardamme, dix-neuf présentent ce poli particulier (Beugnier et Crombé 2007). Cinq ne présentent aucune trace d'utilisation et pourraient constituer un outillage de réserve et

deux ont raclé de la peau sèche et incisé une matière minérale et une matière indéterminée. Le microdenticulé à Waardamme est ainsi clairement dévolu au traitement d'un produit végétal.

Dans ce site, en plus des microdenticulés, neuf éclats bruts présentent des usures identiques. Toutes ces pièces, aux bords relativement fins et donc fragiles, sont en général fortement usées. Les indentations des microdenticulés ne sont plus visibles qu'à l'état de vestige et les bords utilisés bruts de débitage sont marqués par des séries de micro-esquillements de répartition unifaciale. Pour tous ces objets, la face retouchée et la face esquillée correspondent systématiquement à la face en dépouille, autrement dit, la face en contact avec la matière travaillée. Sur la face opposée, soit la face d'attaque, le poli est extrêmement brillant. Il est généralement visible à l'œil nu. Au microscope, il apparaît sous la forme d'une coalescence hautement caractéristique, unie, légèrement ondulée et dépourvue de marqueurs cinétiques (fig. 9).

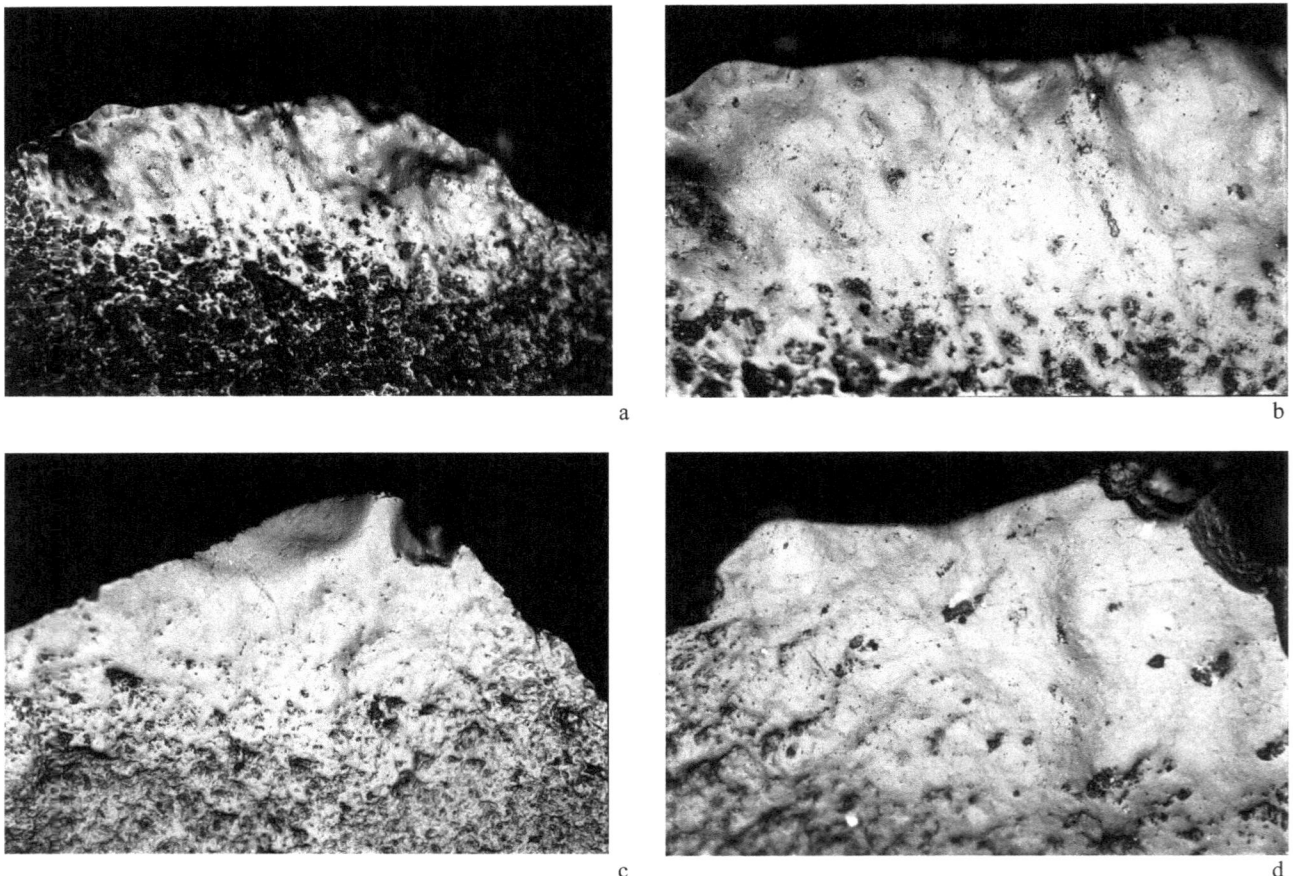

Figure 9. Le poli de face d'attaque uni, mou, brillant, hautement caractéristique des microdenticulés du site de Waardamme : a et b. Microdenticulé n° 149 (X100, X200) ; c. Microdenticulé n° 166 (X200) ; d. Microdenticulé n° 124 (X200).

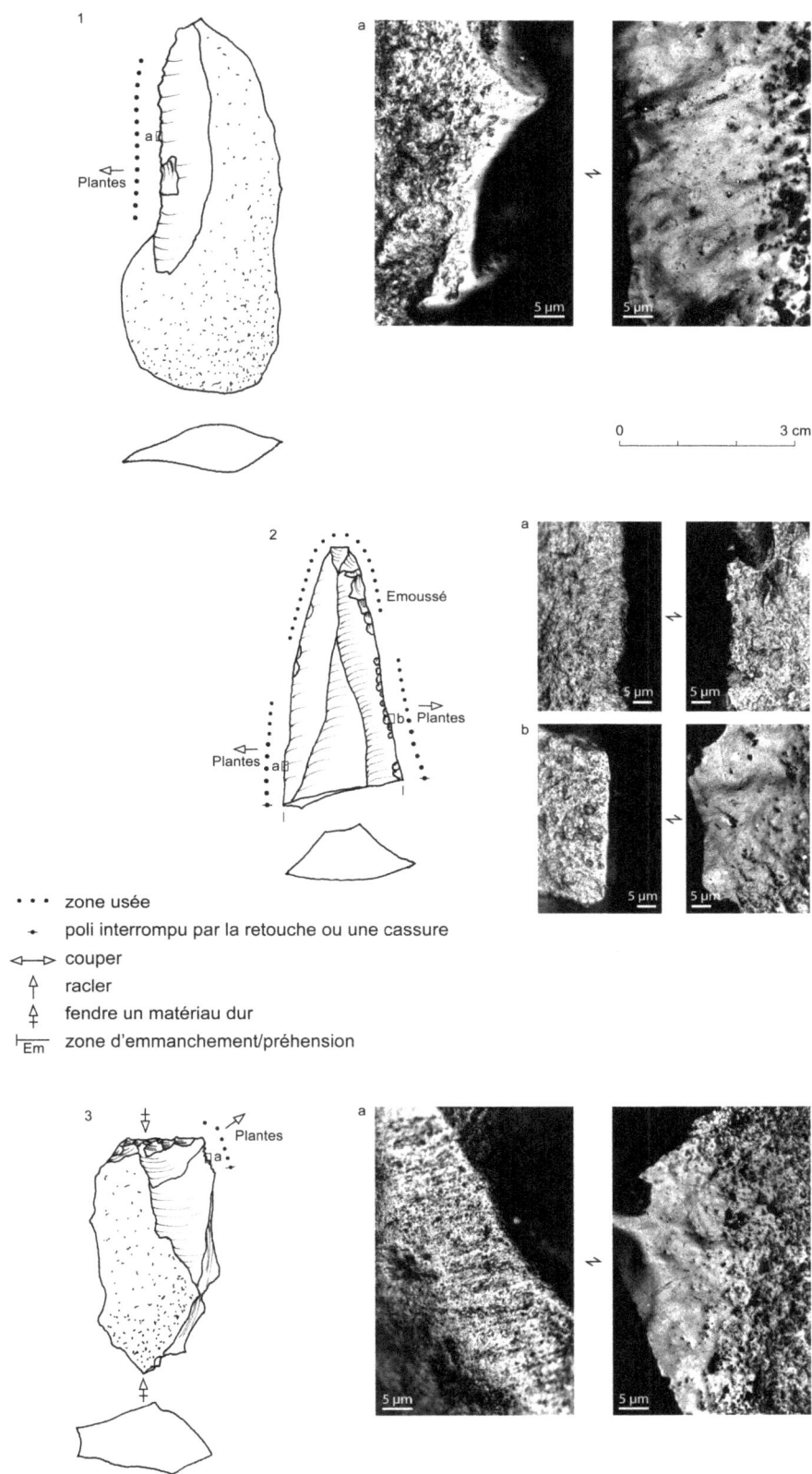

Figure 10. Les polis de raclage observés sur les microdenticulés du site de Waardamme. Variations des traces d'usure de la face en dépouille : 1. Éclat laminaire microdenticulé. À gauche, poli de face en dépouille observé sur la face inférieure. À droite, poli uni, mou, brillant, caractéristique de la face d'attaque (X200) ; 2. Fragment de lame microdenticulée. a : à gauche, poli de face en dépouille observé sur la face inférieure ; à droite, poli de face d'attaque ; b : à gauche, poli de face en dépouille observé sur la face supérieure ; à droite, poli uni, mou, brillant caractéristique de la face d'attaque (X200) ; 3. Microdenticulé recyclé en coin à fendre. À gauche, poli de face en dépouille observé sur la face inférieure. À droite, poli uni, mou, brillant de la face d'attaque (X 200).

Sur la face en dépouille, le poli est beaucoup plus ténu et présente une plus grande variabilité d'aspect d'une pièce à l'autre, mais aussi sur une même portion de bord actif (fig. 10). Les traces peuvent ainsi se présenter sous la forme d'un biseau plat uni assez brillant, débordant plus ou moins vers l'intérieur de la pièce, affecté de stries perpendiculaires à légèrement obliques par rapport au bord. Une deuxième version s'observe sous la forme d'un émoussé prononcé du fil du tranchant et d'un poli mat grenu, abondamment strié, semblable à un poli de peau sèche. Sur de nombreuses pièces enfin, les deux formes d'usure peuvent coexister. Le tranchant est alors marqué par un biseau uni dur, se transformant, vers l'intérieur de la pièce, en un poli mat grenu, strié, émoussant les zones en relief. Ce double aspect des polis de la face en dépouille ne paraît pas dépendre d'utilisations différentes des outils. Archéologiquement, cette association de polis est fréquente. On l'observe, en effet, dans des contextes variés. Par exemple, les importantes séries de microdenticulés des sites danois Ertebølle et TRBK ont fourni toute la variété de stigmates décrite précédemment (Juel Jensen 1994). Dans le nord de la France, les mêmes observations ont été réalisées sur les microdenticulés des sites de Raillencourt-Sainte-Olle et d'Annœullin (Beugnier 2000, 2001, Martial et al. 2004). Par ailleurs, suite aux expérimentations menées, on sait qu'une même opération technique peut produire des traces d'aspect distinct, simplement en modifiant légèrement l'angle entre l'outil et la matière travaillée. La présence ou non de particules de terre, souillant le matériau mis en œuvre, est aussi un facteur de variabilité des traces d'usage. Il est ainsi fort probable que les différentes formes d'usure observées sur la face en dépouille des microdenticulés résultent d'une seule et même utilisation. À Waardamme, un seul outil présente, au niveau du bord opposé à la zone active, des usures ténues, interprétées comme des stigmates de préhension à main nue. À cette exception près, aucune trace de préhension ou d'emmanchement n'a été repérée sur ces outils.

Toutes les observations faites ici recoupent les résultats obtenus sur d'autres ensembles de microdenticulés, quel que soit leur contexte de découverte. Le microdenticulé est présent de la fin du Mésolithique à la fin du Néolithique, où il apparaît et disparaît, sous la forme de grandes séries ou de façon isolée, au sein d'une large zone géographique qui s'étend du Danemark à la moitié nord de la France. Les premiers microdenticulés proviennent de sites danois Ertebølle et TRBK où ils représentent, jusqu'au Néolithique ancien,

entre 2 % et jusqu'à 22 % de l'outillage retouché. Dans cette région, ils disparaissent ensuite complètement des inventaires (Juel Jensen 1994). En Grande-Bretagne, ils se retrouvent par centaine dans les sites enclos du Néolithique ancien et moyen puis décroissent de façon remarquable au Néolithique final et à l'Âge du Bronze (Saville 2002, Hurcombe ce volume) alors que dans la moitié nord de la France et de façon plus limitée en Belgique, les microdenticulés sont strictement liés au Néolithique final, période durant laquelle leur représentation va varier de façon majeure d'un groupe régional à l'autre. Précisément, ils se développent entre 2500 et 2200 av. J.-C. et occupent une place majeure, parfois dominante, dans des assemblages du Néolithique récent du Centre-Ouest (Burnez 1976, Burnez et Fouéré 1999), de la Civilisation Saône-Rhône comme à Charavines (Isère) (Bocquet 1980), du groupe de Chalain dans le Jura (Pétrequin et Pétrequin 1988) et du Deûle/Escaut dans le nord (Blanchet 1984, Piningre 1985, Martial et al. 2004).

En revanche, dans les sites du Gord par exemple, groupe culturel cousin du Deûle/Escaut, ils sont inexistants ou ne sont représentés que par quelques pièces isolées (par exemple, Augereau 1996). Une analyse chronologique fine, menée auprès de sites de la région des lacs dans l'est de la France, a également permis de mettre en évidence une évolution de ces outils très rapide. « Jusque vers 2650 av. J.-C. les éclats de silex microdenticulés sont complètement inconnus en Suisse occidentale et dans le Jura méridional. À Clairvaux (Jura) vers 2600, on en compte à peine 2 % sur le total des outils en silex ; vers 2500, les microdenticulés forment déjà 25 % de l'outillage et en 2400, ils atteignent le chiffre record de 40 % de l'industrie lithique taillée. À Charavines sur le lac de Paladru (Isère), en quelques décennies, la progression est du même ordre. Une telle vitesse d'expansion d'un outil ou d'une technique est absolument unique dans le Néolithique final » (Pétrequin et Pétrequin 1988, p. 238-239). Par ailleurs, comme en atteste aujourd'hui un nombre assez conséquent d'observations tracéologiques, tous ces microdenticulés, quelles que soient leur origine géographique et leur appartenance culturelle, présentent les mêmes usures de raclage dues au contact avec une matière végétale siliceuse indéterminée (Vaughan et Bocquet 1987, Juel Jensen 1994, Beugnier 2000, 2001, inédit a, b, c, Beugnier et Crombé 2007, Hurcombe ce volume, Plisson inédit) (fig. 11).

Ces instruments, faciles à fabriquer, constituent ainsi des éléments extrêmement spécialisés, liés à

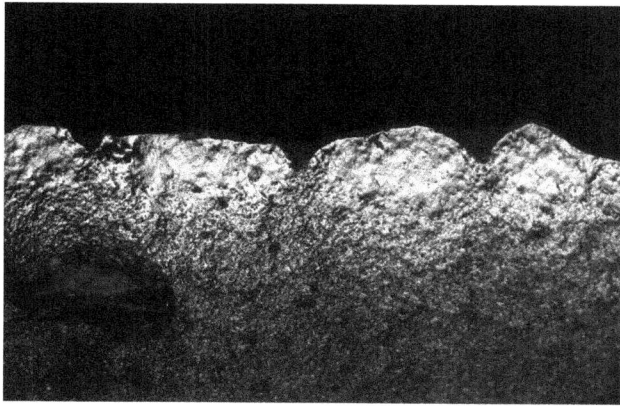
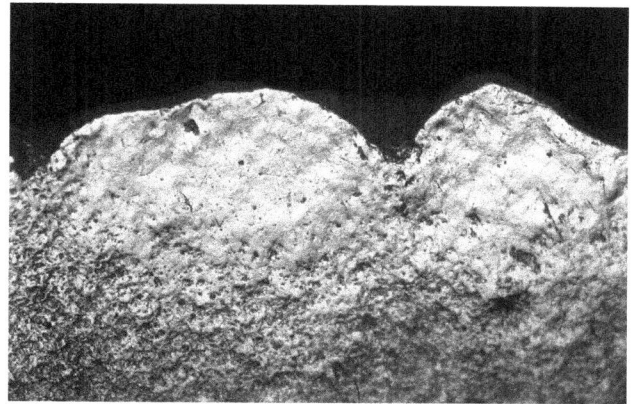

a

b

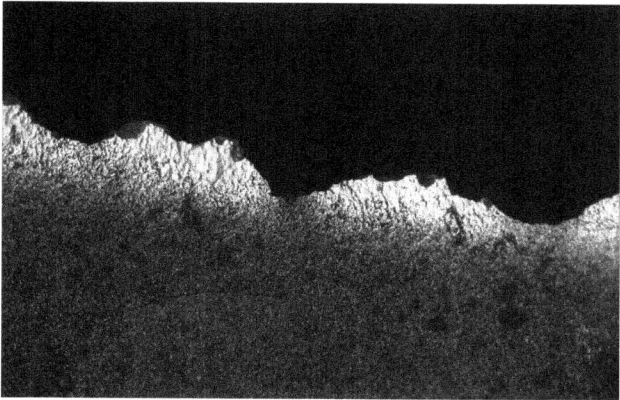

c

d

Figure 11. Détails des polis de la face d'attaque, observés sur divers microdenticulés provenant de sites datés de la fin du Néolithique final : a et b. Site d'Annœullin (Nord) (X50, X100) (Beugnier 2000) ; c et d. Site de Raillencourt-Sainte-Olle (Nord) (X50, X100) (Beugnier 2001).

une activité artisanale ou alimentaire. Ils ne sont utilisés en grand nombre que par certains groupes culturels, d'autres les refusant (au profit d'autres catégories typologiques d'outils ?) ou ne les employant que de façon tout à fait anecdotique. Par ailleurs, quand ces outils disparaissent, ils ne semblent pas être remplacés, du moins par un autre type d'outils en silex. On est donc là face à un phénomène assez mystérieux de diffusion d'une technique et d'un outillage qui va durer presque deux millénaires. La présence des microdenticulés suscite ainsi pas mal d'interrogations à l'origine d'un certain nombre de recherches notamment expérimentales visant à établir la fonction de ces instruments.

Différentes matières végétales utilisées au Néolithique pour la fabrication d'objets variés tels que les nasses, les cordes, les tissus et les hampes de flèche ont ainsi été testées. H. Juel Jensen (1988) a écorcé et raclé des tiges et du liber de saule (*Salix*) et de tilleul (*Tilia*) ; A. van Gijn (1989), des branches de bois, des tiges de lin (*Linum*

Usitatissimum L.) putréfiés et des ronces (*Rubus*) ; H. Plisson (inédit), de l'écorce de bouleau (*Betula*) trempé ; B. Gassin (1996), P. Pétrequin (com. pers.) et J.-P. Caspar (Caspar *et al.*, 2005), des cannes de Provence (*Arundo donax*), des roseaux (*Phragmites australis*) (fig. 12) et des tiges de plantes fibreuses traitées par rouissage. P. Anderson (sous-presse) a obtenu, quant à elle, d'assez bons résultats lors d'opérations de moisson d'épeautre (*Triticum spelta*), réalisées par étêtage des épis mûrs. Cette expérimentation a engendré, sur une des faces des outils, un poli uni, mou, extrêmement brillant, assez proche de celui marquant les faces d'attaque des microdenticulés néolithiques (fig. 13). En revanche, elle n'a pas permis de reproduire la variabilité des usures observées sur les faces en dépouille des pièces archéologiques. Avec ce résultat se pose également un problème concernant l'activité identifiée. En effet, comment valider une hypothèse proposant l'utilisation d'outils mésolithiques Ertebølle pour la moisson d'épis d'épeautre ? Récemment, un important programme expérimental, présenté en

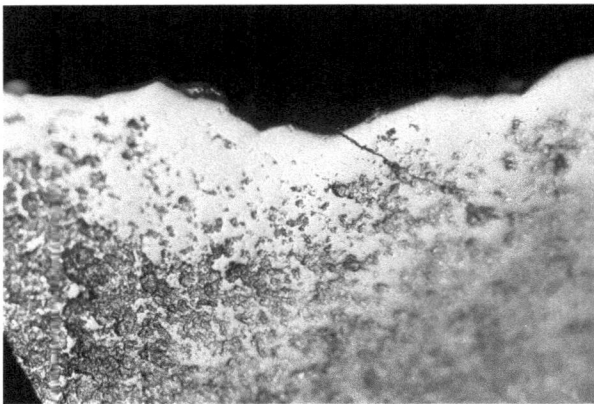

Figure 12. Détails d'usures expérimentales produites lors du raclage de tiges de roseaux (*Phragmites australis*) à l'aide de microdenticulés. En haut, face d'attaque ; en bas, face en dépouille (X200).

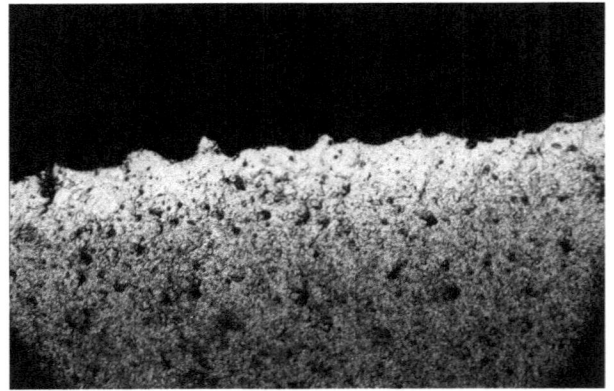

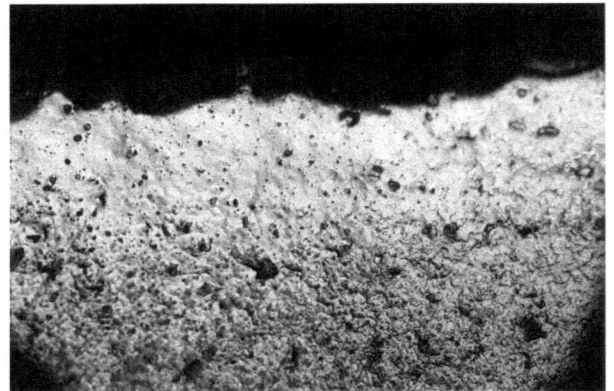

Figure 13. Détails du poli expérimental produit par P. Anderson (sous presse) par l'étêtage d'épis mûrs d'épeautre (X100, X200).

détail dans ce volume, a également été mené sous la direction de L. Hurcombe (ce volume), centré sur la question des micro-denticulés et le traitement des fibres végétales servant à fabriquer textiles et cordages. Alors que différents matériaux (fibres d'ortie, d'écorces de tilleul et de saule, brutes, rouies ou trempées dans des bains de cendre) et modes opératoires ont été testés, les résultats les plus prometteurs ont été obtenus en raclant, de façon prolongée (6 heures), l'épiderme de tiges d'orties (*Urtica dioica*).

Selon l'auteur, le programme doit toutefois être poursuivi, différentes variables (durée d'utilisation, saison, âge, conditions de développement des plantes, etc.) restant à tester qui permettront peut-être d'expliquer l'écart existant toujours entre usure expérimentale et usure archéologique. Cela reste le premier programme expérimental de cette importance, combinant données archéologiques, ethnographiques et l'expertise d'artisans actuels, dans le cadre de chaînes opératoires réalistes aboutissant à un produit fini de qualité.

En regard des échecs passés, les résultats qui ont été obtenus ici sont, pour nous, la preuve que ce type de programme représente la voie à suivre.

CONCLUSION

À l'image de la plupart des recherches présentées dans ce volume (Hardy, Harris, Hurcombe, Martial et Médard), l'étude fonctionnelle des sites de Verrebroek, Doel et Waardamme a permis de restituer toute l'importance du travail des plantes au sein de différentes communautés du Mésolithique et de la fin du Néolithique. Dans les trois sites étudiés, le raclage des matières végétales constitue, en effet, la fonction dominante des outillages en silex.

À Verrebroek et Doel, datés du Mésolithique ancien, ces résultats représentent les premiers témoins d'une exploitation massive des végétaux à l'aide d'outils en silex. S'agissant de sites d'habitat temporaire, « spécialisés » autour des activités cynégétiques, la prédominance des traces d'utilisation dues au contact avec les matières végétales est particulièrement significative, nous rappelant que, pour certains groupes, travailler les plantes a constitué une activité quotidienne permanente. Dans le cadre de cette étude, il n'a malheureusement pas été possible d'établir précisément quelle opération a été réalisée avec ces outils, aucune des expériences menées n'ayant permis de reproduire les usures archéologiques observées. Les recherches les plus récentes, en partie fondées sur l'ethnographie (par exemple, Hardy ce volume), suggèrent de nouvelles hypothèses en rapport notamment avec les chaînes opératoires d'extraction et de production des fibres utilisées pour fabriquer fils, ficelles et cordes. Ces hypothèses restent à valider par l'expérimentation, des résultats prometteurs ayant d'ores et déjà été obtenus par L. Hurcombe (ce volume).

À Waardamme, site Deûle/Escaut de la fin du Néolithique, le travail des plantes est essentiellement représenté par les microdenticulés, marqués, comme partout ailleurs, par un poli de raclage caractéristique dû au contact avec une matière végétale siliceuse indéterminée. Cette étude a ainsi permis d'étendre l'aire de répartition de ces outils, au Néolithique final, à la zone des sols sableux du nord de la Belgique. Elle a aussi été l'occasion de démontrer que de simples éclats bruts ont pu être utilisés, parallèlement aux microdenticulés, pour la même tâche. Ces produits bruts n'étant habituellement pas identifiés dans les sites, l'importance de l'activité menée a donc probablement été systématiquement sous-estimée. Dans ce cas comme précédemment, l'absence d'usure expérimentale identique aux stigmates observés sur le matériel archéologique ne permet aucune conclusion sur la fonction précise de ces outils. Des expérimentations sont toujours en cours, certaines en rapport avec les chaînes opératoires de production de fibres d'orties et d'écorces d'arbre utilisées pour la fabrication des cordages et des textiles (Hurcombe ce volume).

Cet article a ainsi permis de mettre évidence certains des problèmes rencontrés lors de l'interprétation en termes d'activité des traces d'utilisation liées aux matières végétales. Il a aussi permis de montrer l'importance des collaborations entre disciplines complémentaires. L'élaboration des hypothèses comme la réalisation des expérimentations demandant un tel niveau de connaissances techniques et de savoir-faire qu'on ne peut plus faire, dans ce domaine, l'économie de recherches pluridisciplinaires utilisant les compétences des ethnologues, des ethno-historiens, des botanistes et des artisans.

REMERCIEMENTS

Les recherches présentées ont été menées dans le cadre de différents projets, dirigés par Ph. Crombé et financés par le fonds de recherche (Bijzonder Onderzoeksfonds) de l'Université de Gand (Belgique), en collaboration avec l'Institut royal des Sciences naturelles de Belgique.

ADRESSE DE L'AUTEUR

Valérie Beugnier
Section Anthropologie et Préhistoire
Institut royal des Sciences naturelles de Belgique
Rue Vautier, 29
B-1000 Bruxelles – Belgique
valerie.beugnier@yahoo.fr

BIBLIOGRAPHIE

ANDERSEN, S. H., 1987. Tybrind Vig: a submerged Ertebølle Settlement in Denmark. *In*: J. M. Coles and A. J. Lawson, eds. *European Wetlands in Prehistory*. Oxford: University Press, 253-280.

ANDERSON, P., éd., 1992. *Préhistoire de l'agriculture*. Monographie du CRA, 6. Paris : CNRS Éditions.

ANDERSON, P., sous-presse. Plant processing traces in early agriculture sites in the Middle East, ethnoarchaeology and experiments. *In*: A. S. Fairbairn and E. Weiss, eds. *Ethnobotanist of distant past: Archaeobotanical studies in Honour of Gordon Hillman*.

ANDERSON-GERFAUD, P., 1981. *Contribution méthodologique à l'analyse des microtraces d'utilisation sur les outils préhistoriques*. Thèse de 3e cycle. Université de Bordeaux I.

ASTRUC, L., 2002. *L'outillage lithique taillé de Khirokitia. Analyse fonctionnelle et spatiale*. Monographie du CRA, 25. Paris : CNRS Éditions.

AUGEREAU, A., 1996. Le site néolithique final de Bazoches-lès-Bray/Le Tureau à l'Oseille (Seine-et-Marne). *Internéo*, 1, 127-139.

BATS, M., CROMBÉ, Ph., PERDAEN, Y., SERGANT, J., VAN ROEYEN, J.-P. en VAN STRYDONCK, M., 2003. Nieuwe ontdekkingen in het Deurganckdok te Doel (Beveren, Oost-Vlaanderen) : Vroeg- en Finaal-Mesolithicum. *Notae Prahistorica*, 23, 55-59.

BEUGNIER, V., 2000. Étude fonctionnelle des microdenticulés, des tranchets et des racloirs à coches : rapport préliminaire. *In* : I. Praud, éd. *Des occupations mésolithique et néolithique à*

Armoeullin « rue Lavoisier » (Zone 1), Rapport de fouille. SRA du Nord-Pas-de-Calais (inédit), 82-85.

BEUGNIER, V., 2001. Étude fonctionnelle des microdenticulés du site de Raillencourt-Sainte-Olle (Nord). *In* : E. Martial, ed. *Raillencourt-Sainte-Olle « Le Grand Camp », ZAC Actipôle de l'A2. Rapport de fouille*. SRA du Nord-Pas-de-Calais (inédit).

BEUGNIER, V., 2006a. Étude tracéologique du matériel des concentrations C22, C33, C34, C35 et C59 et de 20 pièces hors concentration du site de Verrebroek (Belgique). Rapports inédits. Université de Gand.

BEUGNIER, V., 2006b. Étude tracéologique du matériel mésolithique de Doel (Belgique). Rapport inédit. Université de Gand.

BEUGNIER, V., 2007. Étude tracéologique du matériel swifterbant du site de Doel, secteur B (Belgique). Rapport inédit. Université de Gand.

BEUGNIER, V., inédit a. Château-Landon « Le camp » (Seine-et-Marne). Analyse fonctionnelle de l'industrie en silex. Rapport inédit. Institut National de Recherches Archéologiques et du Patrimoine.

BEUGNIER, V., inédit b. Le site du Néolithique final de Bazoche-lès-Bray (Seine-et-Marne). Évaluation tracéologique du matériel en silex taillé. Rapport inédit. Institut National de Recherches Archéologiques et du Patrimoine.

BEUGNIER, V., inédit c. Étude tracéologique d'outils en silex pressigniens du Petit-Paulmy (Abilly, Indre-et-Loire). Rapport inédit.

BEUGNIER, V. et CROMBÉ, Ph., 2005. Étude fonctionnelle du matériel en silex du site mésolithique ancien de Verrebroek (Flandres, Belgique) : premiers résultats. *Bulletin de la Société préhistorique française*, 102 (3), 527-538.

BEUGNIER, V. et CROMBÉ, Ph., 2007. L'outillage commun du premier site d'habitat découvert en Flandre (Belgique). Étude fonctionnelle de l'industrie lithique de Waardamme (3ᵉ millénaire av. J.-C.). *Bulletin de la Société préhistorique française*, 104 (3), 525-542.

BIENENFELD, B. L. , 1985. Preliminary Results from a Lithic Use-wear Study of Swifterbant Sites S-51, S-4 and S-2. *Helinium*, 25, 194-211.

BLANCHET, J.-C., 1984. *Les premiers métallurgistes en Picardie et dans le nord de la France. Chalcolithique, âge du Bronze et premier âge du Fer*. Mémoires de la Société préhistorique française, 17. Paris : Société Préhistorique Française.

BOCQUET, A., 1980. Les microdenticulés, un outil mal connu. *Bulletin de la Société préhistorique française*, 77, 7-17.

BURNEZ, C., 1976. *Le Néolithique et le Chalcolithique dans le Centre Ouest de la France*. Mémoire de la Société préhistorique française, 12. Paris : Société Préhistorique Française.

BURNEZ, C. et FOUÉRÉ, P., 1999. *Les enceintes néolithiques de Diconche à Saintes (Charente-Maritime). Une périodisation de l'Artenac*. Mémoire de la Société préhistorique française, 25. Paris : Société Préhistorique Française.

CASPAR, J.-P., FÉRAY, Ph. et MARTIAL, E., 2005. Identification et reconstitution des traces de teillage des fibres végétales au Néolithique. *Bulletin de la Société préhistorique française*, 102 (4), 867-880.

CLÉMENTE CONTE, I., 1995. *Instrumentos de trabajo liticos de los yamanas (Canoeros-nomadas de la Tierra des Fuego) : una perspectiva desde el analisis funcional*. Tesis doctoral. Universitat Autonoma de Barcelona.

COLES, B. and COLES, J., 1989. *People of the Wetlands. Bogs, Bodies and Lake-Dwellers*. London: Thames and Hudson.

COLES, B. and LAWSON, A. J., eds, 1987. *European Wetlands in Prehistory*. Oxford: Clarendon Press.

CROMBÉ, Ph., 1998. *The Mesolithic in North-western Belgium, Recent excavations and surveys*. BAR International Series 716. Oxford: Archeopress.

CROMBÉ, Ph., ed., 2005. *The Last Hunter-Gatherer-Fishermen in Sandy Flanders (NW Belgium). The Verrebroek and Doel Excavation Projects (Vol. 1)*. Archaeological Reports Ghent University, 3. Ghent: Academia Press.

CROMBÉ, Ph., PERDAEN, Y., SERGANT, J. and CASPAR, J.-P., 2001. Wear Analysis on Early Mesolithic Microliths from the Verrebroek site, East Flanders, Belgium. *Journal of Field Archaeology*, 28 (3-4), 253-269.

CROMBÉ, Ph., PERDAEN, Y. and SERGANT, J., 2003. The Site of Verrebroek Dok (Flanders, Belgium): Spatial Organisation of an Extensive Early Mesolithic Settlement. *In*: L. Larsson, H. Kindgren, K. Knutsson, D. Leoffler and A. Akerlund, eds. *Mesolithic on the Move: Papers presented at the Sixth International Conference on the Mesolithic in Europe, Stockholm 2000*. Oxford: Oxbow Books Ltd., 205-215.

DEFORCE, K., GELORINI, V., VERBRUGGEN, C. and VRYDAGHS, L., 2005. Pollen and phytolith analyses. *In*: Ph. Crombé, ed. *The Last Hunter-Gatherer-Fishermen in Sandy Flanders (NW Belgium). The Verrebroek and Doel Excavation Projects (Vol. 1)*. Archaeological Reports Ghent University, 3. Ghent: Academia Press, 108-126.

DEMEYERE, F., BOURGEOIS, J., CROMBÉ, Ph., VAN STRYDONCK, M., 2006. New Evidence of the Final

Neolithic Occupation of the Sandy Lowland of Belgium: the Waardamme « Vijvers » site, West Flanders. *Archäologisches Korrespondenzblatt*, 36 (2), 179-194.

GASSIN, B., 1996. *Évolution socio-économique dans le Chasséen de la grotte de l'Église supérieure (Var). Apport de l'analyse fonctionnelle des industries lithiques.* Monographie du CRA, 17. Paris : CNRS Édition.

GONZALEZ URQUIJO, J. E., IBANEZ ESTEVEZ, J. J., 1994. *Metodogia de analisis funcional de instrumentos tallados en silex.* Cuadernos de Arqueologia, 14. Bilbao : Universitad de Deusto.

GRACE, R., 1989. *Interpreting the Function of Stone Tools. The quantification and computerisation of microwear analysis.* BAR International Series 474. Oxford: Archeopress.

GRAMSCH, B., 1987. Ausgrabungen auf dem mesolithischen Moorfundplatz bei Friesack, Bezirk Postdam. *Veröffentlichungen des Museums für Ur-und Frühgeschichte Postdam*, 21, 75-100.

GROENEWOUDT, B.J., DEEBEN, J., VAN GEEL, B. and LAUWERIER, R.C.G.M., 2001. An Early Mesolithic Assemblage with Faunal Remains in a Stream Valley near Zutphen, the Netherlands. *Archäologisches Korrespondenzblatt*, 31, 329-348.

IBANEZ ESTEVEZ , J. J. and GONZALEZ URQUIJO, J. E., 1996. *From tools use to site function. Use-wear analysis in some upper Palaeolithic sites in the Basque country.* BAR International Series 658. Oxford: Archeopress.

JUEL JENSEN, H., 1988. Microdenticulates in the Danish Stone Age: A functional puzzle. *In*: S. Beyries, ed. *Industries Lithiques, Tracéologie et Technologie, 1, Aspects archéologiques.* BAR International Series 411 (I). Oxford: Archeopress, 231-252.

JUEL JENSEN, H., 1994. *Flint Tools and Plant Working. Hidden Traces of Stone Age Technology. A use wear study of some Danish Mesolithic and TRB implements.* Aarhus: Aarhus University Press.

JUEL JENSEN, H. and BRINCH PETERSEN, E., 1985. A Functional Study of Lithics from Vænget Nord, a Mesolithic Site at Vedbæk, N.E. Sjælland. *Journal of Danish Archaeology*, 4, 40-51.

KÖBER-GROHNE, U., 1995. Bericht über die botanisch-mikroskopische Bestimmung des Rohmaterial von einigen Schnüren, Seilen und Netzen von Friesack, Landkreis Havelland. *Veröffentlichungen des Brandenburgischen Landesmuseums für Ur- und Frühgeschichte*, 29, 7-12.

KOOISTRA, L., 2006. Fabrics of fibres and strips of bark. *In*: L.P. Louwe Kooijmans and P.F.B. Jongste, eds. *Schipluiden: a Neolithic settlement on the Dutch North Sea coast c.3500 CAL BC.* Analecta Praehistorica Leidensia, 37/38. Leiden: Leiden University, 253-259.

LARSSON, L., 1983. *Ageröd V. An Atlantic Bog Site in Central Scania.* Acta Archaeologica Lundensia Series 8, 12. Bonn: Rudolf Habelt Verlag.

LOUWE KOOIJMANS, L. P., 1987. Neolithic Settlement and Subsistence in the Wetlands of the Rhine/Meuse delta of the Netherlands. *In:* J. M. Coles and A. J. Lawson, eds. *European Wetlands in Prehistory.* Oxford: University Press, 227-251.

MARTIAL, E., PRAUD, I., BOSTYN, F., 2004. Recherches récentes sur le Néolithique final dans le nord de la France. *In* : M. Vander Linden et L. Salanova, eds. *Le troisième millénaire dans le nord de la France et en Belgique.* Mémoire de la Société Préhistorique Française, 35, Anthropologica et Praehistorica, 115. Bruxelles : Société royale belge d'Anthropologie et de Préhistoire, 49-72.

MEERSSCHAERT, L., 2005. Botanical macroremains : seeds and fruits. *In:* Ph. Crombé, ed. *The Last Hunter-Gatherer-Fishermen in Sandy Flanders (NW Belgium). The Verrebroek and Doel Excavation Projects (Vol. 1).* Archaeological Reports Ghent University, 3. Ghent: Academia Press, 261-266.

PEETERS, J.H.M., SCHREURS, J. en VERNEAU, S.M.I.P., 2001. Deel 18. Vuursteen: typologie, technologische organisatie en gebruik. *In:* J.W.H. Hogestijn en J.H.M Peeters, eds. *De mesolithische en vroeg-neolithische vindplaats Hoge Vaart-A27 (Flevoland).* Rapportage Archeologische Monumentenzorg 79. Amersfoort: Rijksdienst voor het Oudheidkundig Bodemonderzoek.

PERRY, D., 1999. Vegetative Tissues from Mesolithic Sites in the Northern Netherlands. *Current Anthropology*, 40 (2), 231-237.

PÉTREQUIN, P., 1984. *Gens de l'Eau, Gens de la Terre, ethno-archéologie des communautés lacustres.* Paris : Hachette littérature.

PÉTREQUIN, P., ed., 1986. *Les sites littoraux néolithiques de Clairvaux-les-Lacs (Jura), 1, Problématique générale, L'exemple de la station III.* Paris : Éd. Maison des Sciences de l'Homme.

PÉTREQUIN, P., éd., 1989. *Les sites littoraux néolithiques de Clairvaux-les-Lacs (Jura), II, le Néolithique moyen.* Paris : Éd. Maison des Sciences de l'Homme.

PÉTREQUIN, P., éd., 1997. *Les sites littoraux néolithiques de Clairvaux et Chalain (Jura), III, Chalain station 3, 3200-2900 av. J.-C.* Paris : Éd. Maison des Sciences de l'Homme.

PÉTREQUIN, P. et PÉTREQUIN, A.-M., 1988. *Le Néolithique des lacs. Préhistoire des lacs de Chalain et de Clairvaux*. Paris : Éditions Errance.

PHILIBERT, S., 2002 . *Les Derniers « Sauvages ». Territoires économiques et systèmes techno-fonctionnels mésolithiques*. BAR International Series 1069. Oxford: Archeopress.

PININGRE, J.-F., 1985. Un aspect de la fin du Néolithique dans le Nord de la France. Les sites de Seclin, Houplin-Ancoisne et Saint-Saulve (Nord). *In* : *Le Néolithique dans le nord de la France et le Bassin parisien. Actes du 9ᵉ colloque interrégional sur le Néolithique*. Revue Archéologique de Picardie, 3-4. Amiens : Société des Antiquités Historiques de Picardie, 53-69.

PLISSON, H., 1985. *Étude fonctionnelle d'outillages lithiques préhistoriques par l'analyse des micro-usures : recherche méthodologique et archéologique*. Thèse de 3ᵉ cycle. Université de Paris I.

PLISSON, H., inédit. Analyse tracéologique d'une série de microdenticulés du site de Charavines (Isère). Rapport du Centre de Documentation de la Préhistoire Alpine, Grenoble.

SAVILLE, A., 2002. Lithic Artefacts from Neolithic Causewayed Enclosures: Character and Meaning. Lithic studies. *In*: G. Varndell and P. Topping, eds. *Enclosures in Neolithic Europe. Essays on Causewayed and Non-Causeweyed Sites*. Oxford: Oxbow Books, 91-105.

SKAARUP, J., 1983. Submarine stenalderbopladser i det sydfynskeohav'. *Antikvariske studier*, 6, 137-161.

VAN GIJN, A. L., 1989. *The wear and tear of flint. Principles of functional analysis applied to Dutch Neolithic assemblages*. Analecta Praehistorica Leidensia, 22. Leiden: University of Leiden.

VAN GIJN, A. L., BEUGNIER, V. en LAMMERS-KEIJSERS, Y., 2001a. Vuursteen. *In*: L. P. Louwe Kooijmans, ed. *Archeologie in de Betuweroute, Hardinxveld-Giessendam Polderweg. Een mesolithisch jachtkamp in het rivierengebied (5500-5000 v. Chr.)*. Rapportage Archeologische Monumentenzorg, 83. Amersfoort: Rijksdienst voor het Oudheidkundig Bodemonderzoek, 119-162.

VAN GIJN, A.L., LAMMERS-KEIJSERS, Y. en HOUKES, R., 2001b. Vuursteen. *In*: L. P. Louwe Kooijmans, ed. *Archeologie in de Betuweroute. Hardinxveld-Giessendam De Bruin. Een kampplaats uit het Laat-Mesolithicum en het begin van de Swifterbant-cultuur (5500-4450 v. Chr.)*. Rapportage Archeologische Monumentenzorg, 88. Amersfoort: Rijksdienst voor het Oudheidkundig Bodemonderzoek, 153-191.

VAN STRYDONCK, M, CROMBÉ, Ph. and MAES, A., 2001. The Site of Verrebroek « Dok » and its Contribution to the Absolute Dating of the Mesolithic in the Low Countries. *Radiocarbon,* 43 (2B), 997-1005.

VAUGHAN, P. C., 1985. *Use-Wear Analysis of Flaked Stone Tools*. Tucson: The University of Arizona Press.

VAUGHAN, P. C. et BOCQUET, A., 1987. Première étude fonctionnelle d'outils lithiques néolithiques du village de Charavines, Isère. *L'Anthropologie (Paris)*, 91 (2), 399-410.

PLANT PROCESSING FOR CORDAGE AND TEXTILES USING SERRATED FLINT EDGES: NEW *CHAÎNES OPÉRATOIRES* SUGGESTED BY COMBINING ETHNOGRAPHIC, ARCHAEOLOGICAL AND EXPERIMENTAL EVIDENCE FOR BAST FIBRE PROCESSING

Linda HURCOMBE

Abstract: Wear analysts have long recognised a particular kind of flint tool prevalent in the late Mesolithic/Neolithic in parts of Europe including Britain and Denmark. The flint tool is distinctive due to small indentations on the working edge which is why it is variously termed a " microdenticulate " or " serrated-edge " flake. It is also characterised by an intense silica-rich polish showing it was used in a scraping motion. Processing plants to produce cordage and fabric are possible solutions to this functional puzzle. Ethnographic evidence and extant remains of cordage and textiles suggested plant processing activities connected with bast fibre production. Experiments have shown what kinds of activities best match the archaeological tool wear traces and the combination of ethnographic, experimental and archaeological evidence has been able to suggest new " chaînes opératoires " for plant processing activities relating to tree bast fibres and nettle bast fibres. Perishable material culture is under-represented in the archaeological record and this research provides ways in which inorganic remains can shed light on the organic materials which were once so prevalent but which survive so rarely.

Keywords: bast fibres, *Urtica dioica*, *Salix*, *Tilia*, usewear, micro-denticulate, serrated, neolithic flint tools.

Résumé : Les tracéologues ont depuis longtemps reconnu un type d'outil particulier présent au Mésolithique final et au Néolithique dans différentes parties d'Europe, telles que la Grande-Bretagne et le Danemark. Cet outil en silex est caractérisé par une série de petites indentations façonnant le bord actif, raison pour laquelle il est diversement désigné par les termes de " microdenticulé " ou " éclat à bord denté ". Il se distingue également par la présence d'un poli intense dû à une utilisation en raclage sur un matériau riche en silice. Le traitement des plantes pour la production de cordes et de textiles constitue une réponse possible au problème fonctionnel que pose cet outil. Les données ethnographiques disponibles ainsi que les fragments conservés de cordes et de textiles, dans les sites archéologiques, suggèrent un rapport entre ces activités de traitement des plantes et la production de filasse (ou fibre de liber). Les expérimentations menées ont montré quel type d'activités produisait les usures les plus proches des traces d'utilisation observées sur le matériel archéologique. La confrontation des données ethnographiques, expérimentales et archéologiques a ainsi permis de suggérer de nouvelles chaînes opératoires du traitement des plantes en rapport avec la production de filasse d'écorce d'arbre et de filasse d'ortie. Alors que les biens périssables sont sous-représentés dans les sites archéologiques, cette recherche fournit, par le biais des matériaux inorganiques, un moyen de documenter les restes organiques autrefois prédominants mais rarement conservés aujourd'hui.

Mots-clés : fibres de liber, ortie (*Urtica dioica*), saule (*Salix)*, tilleul (*Tilia)*, trace d'utilisation, microdenticulé, denté en scie, outils en silex néolithiques.

INTRODUCTION

The research reported here is part of a long term research theme, " organics from inorganics ", which aims to extend our knowledge of the organic material culture which survives so rarely by making more imaginative use of the inorganic artefacts which survive more frequently. A variety of artefact categories and methods are used, in which experimental archaeology, informed by ethnographic data, features strongly. Organic material culture forms the bulk of all possessions and clothes yet it is notoriously under-represented in the archaeological record. Furthermore, the production of fabrics, cordage and textiles is often a female gendered craft activity where the craft skills and *chaînes opératoires* deserve to be documented and accorded more attention for they offer rich possibilities in cultural expression and individual agency (Hurcombe 2000, 2007). These initiatives are born out of interests in gendered craft activities and the need to take a more integrated approach to material culture analyses, and that is why this research is not about a particular local technology but an under-represented, ubiquitous and essential craft sphere. In the summer of 2005 a series of plant harvesting experiments with plain and serrated edges

(Hurcombe in press) were conducted, and experimentation with transverse actions intended to extract the fibres from plants such as nettles (*Urtica dioica*, growing to 1,5 m in the right conditions) and tree bast (e.g. from willow, lime) was begun. The experimental work reported here spans several years and is stimulated by a much wider interest in gender issues and material culture and forms part of a long term project to extend the methodologies available to investigate a neglected field.

The reason for focussing on this topic is that in the Mesolithic of Denmark and elsewhere in Europe there are fragments of " plant fibre " textiles. Indeed, recent work on such topics has shown that the Palaeolithic has a rich tradition of cordage, basketry and textiles (Adovasio *et al.* 1996, Soffer *et al.* 2000, Soffer 2004). Therefore plants are not just *eaten* but *used* to produce complex items of material culture from very early periods onwards.

There is every reason to suppose that the few fragments from the Mesolithic and earlier periods demonstrate an established craft where the creation of fibres must have been a significant component of the range of tasks performed by these people and as such, some aspects of the processing *chaînes opératoires* requiring tools might leave tangible evidence in the form of wear traces on stone or other tools (van Gijn 2006), even though it is accepted that many will involve no tools or ones made only from organic materials (Hurcombe 1998). The few finds which do exist show that bast fibres played an important role (e.g. Andersen 1987, Kooistra 2006). However, there is one example where a tool has been identified with wear traces suggesting it is used in processing plant fibres, but the precise manner of its use has yet to be determined. The flint tools with the distinctive wear traces suggesting transverse plant working occur in a wide area and the task with which they are associated is therefore indicative of a practice that was at one time geographically widespread across temperate Europe but which has a distinct time period. Thus solving the functional puzzle of these distinctive tools is not a particularist study of interest to a few specialists but could reveal an important and widespread aspect of prehistoric life. It could be that as the Neolithic progressed, there was more kemp (coarse hair) and wool from domestic sheep and possibly also a growth in the use of other kinds of plant fibre crops such as flax. In these circumstances the raw materials for textiles and the associated processing techniques could have

changed. Both hemp (once considered a much more recent fibre source but now with evidence for its use in the Bronze Age in Britain, Godwin 1967, Bradshaw *et al.* 1981, Ryder 1999) and flax can be retted and then the flimsy remnants of their woody stems can be removed by using a " flax break " or pounding with a mallet, and such practices also work to a certain extent for nettles. Nettles are common and a neglected possible source of plant fibres (Grieve 1931, Dreyer *et al.* 1996). It was thought that processing them or other bast fibres which are today uncommonly used, might offer some suggestions of the function of a distinctive and prevalent set of stone tools. The fact that the wear traces have not previously been specifically identified to a task and material is not problematic but intriguing. It is accepted that not all activities which were undertaken in the past are known in the present. The wear traces may not have been successfully replicated because of additives (clay, ash?) or special conditions (partially rotted material?) or because the tools were used in a set of activities all related to the same task and so the pattern observed is a palimpsest of related wear traces. Thus solving the puzzle relies on both understanding and, in so doing, revealing the *chaînes opératoires* in which these tools played their role.

SERRATED FLINT EDGES:
A PARTICULAR FUNCTIONAL PUZZLE

H. Juel Jensen (1994) identified the research problem as " *microdenticulates: an unresolved functional puzzle* ". Essentially, it has long been recognised by European wear specialists (van Gijn 1989, Juel Jensen 1994) that there are special tools with a very distinctive set of wear traces which could be due to working plants, and such tools with similar wear traces are known from Neolithic contexts in Britain, but no one has as yet successfully replicated the distinctive wear traces via experiments. The tool motion is across the working edge so they are used for a shallow scraping action. The tool edges are " micro-denticulate " (the term common in Europe) or " serrated " (the more common term in Britain). Figures 1 and 2 show the fine edge scarring in detail. Figure 3 shows the distinctive intense polish on the ventral surface and on the tips of the serrations with the striations and polish arrangement indicating a transverse but light use action. Such intense polishes are normally associated with siliceous materials. Some of the serration tips also have microscopic edge damage

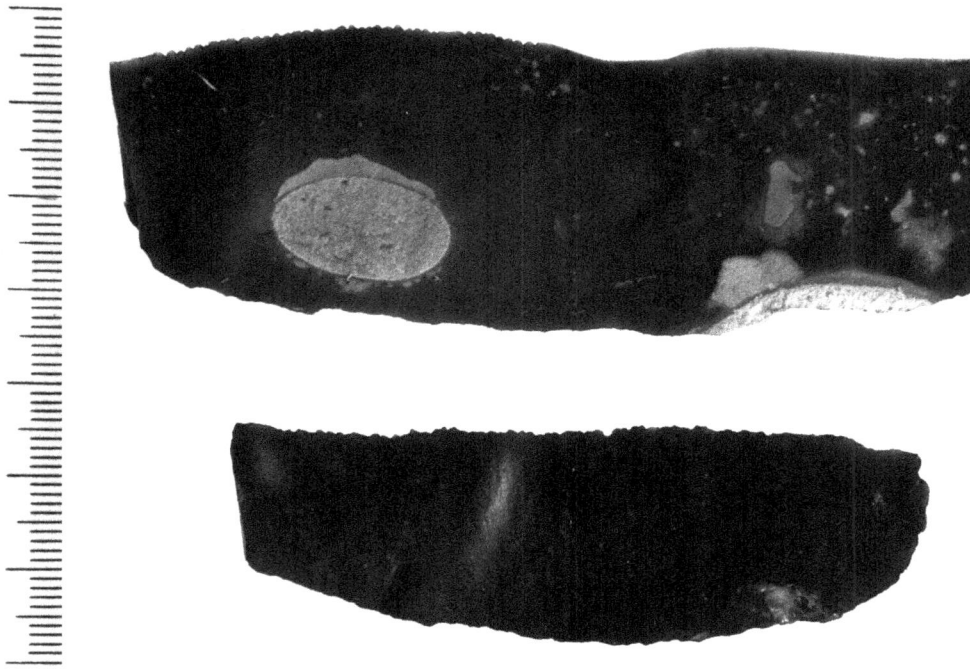

Figure 1. Experimental (top) and archaeological (bottom) serrated tool ventral faces showing micro-denticulated edges.

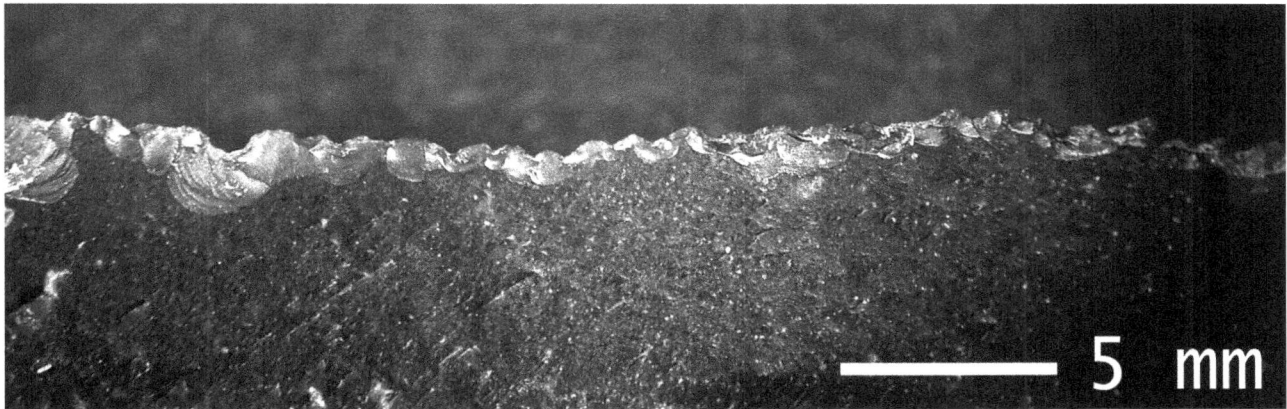

Figure 2. Details of flake scars on dorsal surface of archaeological tool.

(fig. 4). Such small scars associated with the polished areas are not a side effect of the deliberate retouch to create the serrations, but are in keeping with the transverse motion of use and form part of the use traces. It was thought that since the implements could be fibre processing tools, the characteristics and performance of serrated edges should be investigated as part of a longer term experimental programme. However, their period, geographical range, and representation on different kinds of sites, add to this functional puzzle and require discussion.

Serrated edge tools are found on causewayed enclosure sites in a variety of locations such as the wetland site of Etton (Middleton 1998) as well as Abingdon (Case and Whittle 1982), and Briar Hill (Bamford 1985), and from other kinds of enclosures such as Carn Brea, although Mercer (1981) called his tools by a slightly different name. They also occur on a wide range of other sites from the assemblage of the Somerset Levels (Brown 1986) to funerary sites such as at the long cairn at Hazleton North (Saville 1990) and at Newnham Murren ring ditch, the grave pit contained a serrated flake beside the knee of a middle aged female skeleton (Moorey 1982). As with the flint tools from continental Europe mentioned above, the serrated pieces are made on a variety of different shaped blanks, some have

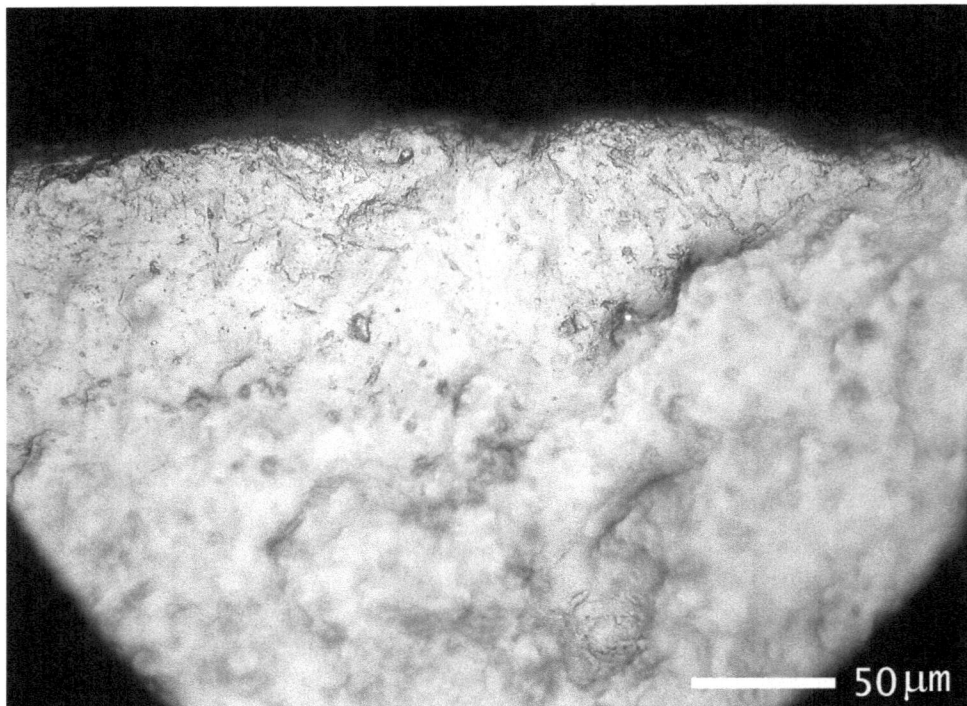

Figure 3. Views of the intense polish and wear traces on the ventral surface of the archaeological tool. Original magnification (OM) 50X, 200X, 500X.

Figure 4. Archaeological tool ventral face to show microscopic edge scarring on the tips of some serrations. OM 50X.

macroscopically visible polish, and some are on concave curved edges. These points have been made in relation to searching for evidence of plant craft-working activities across organic, inorganic and environmental categories of evidence (Hurcombe 2000), and as part of a review of the character and meaning of lithic artefacts from causewayed enclosures (Saville 2002). These tools are part of the assemblages of Neolithic enclosures across North-West Europe from France to Denmark giving them a strong geographical range, and they are also very common tools in most British enclosures, making them numerically a significant category within the site assemblages, yet their time range is quite distinctive as in Britain they are common in early and middle Neolithic contexts but their numbers decline sharply in the late Neolithic (*op cit.*). In Denmark they are known from final Mesolithic periods and the earlier Neolithic (e.g. Ertebølle and TRB contexts in Denmark) and in France from the Neolithic (Bocquet 1994).

Figure 5 draws out the comparative numbers of " types " of lithic implements in some British causewayed enclosures. It is accepted that some of these types are loose categories where individual researcher's classification systems and nomenclature will differ and affect the numbers presented, yet the gross pattern can be seen by comparing the numbers of scrapers against the numbers of serrated-edge flakes. Most researchers recognize a scraper but there is not the same confidence level for serrated flakes yet the numbers for the two classes are not so different. Furthermore, when looking through artefacts in museum collections it

is obvious that many " serrated flakes " have less regular serrations, perhaps due to a loose mental template of the type, less care in the original manufacturing process, a single or series of retouching episodes, or such irregularity could be due to some use damage. Whatever the cause, the effect is to make the " serrated flake " less evident and I have seen some of these less regular items listed as " utilised flake " or " edge-damaged " or " retouched flake ". Thus in figure 5, the " serrated-edge flake " and " edge-trimmed flake " categories are deliberately juxtaposed, because many tools serving the same function will have fallen into this much looser second category (also see van Gijn 1998 for a discussion of functions vs tool types). For example, figure 6 shows some tools from the Neolithic Sweet Track flints from the Somerset Levels with less regular edge scarring but with wear traces which link them to the transverse motion, and silica-rich wear traces of the tools described as " serrated-edge flakes " (Hurcombe in press). Thus the gross pattern evident from the archaeological evidence is nonetheless blurred by the practical difficulties of classification systems suggesting that figure 5 may seriously under-represent the quantities of pieces contributing to the " functional puzzle " of serrated-edge flakes.

One further question leaps out from these tools. To anyone with experience of using stone tools, these serrated edges look vulnerable. The small projections in a brittle material such as flint lead to uneven pressure during use and, if used with very forceful actions, or on very hard materials, any experienced flint-user would know they would be quickly damaged and rendered ineffective.

	HH (ME)	HH (Ste)	Et	WH	BH	Sta	Ab	MC	*Total*	Comments
Piercers	86	49	58	68	34	160	5	17	*477*	Hide working but also other materials.
Scrapers	351	240	253	1399	181	377	166	421	*3388*	Hide working and also wood and bone/antler
Serrated edge flakes[1]	343	146	219	627 +	83	195 +	272	188	*2073+*	Many have macroscopically visible gloss suggesting a siliceous contact material, probably siliceous plant; usually only very regular edge-scar serrations are included in this category.
"Edge-trimmed" flakes	411	184	410	457	-	-	370	-	*3905?*	Likely that some or many of this category have wear similar to that of serrated flakes since some have macroscopic gloss and irregular edge-scarred edges.
Knives	7	4	1	84 +	11	104 ?	10	12	-	
"Laurel leaf" bifacials	3	3	5	25	1	66 ?	5	2	-	
Leaf arrowheads	29	9	29	132	18	34	17	28	-	
"Fabricators"/ rods	7	6	9	2 +	2	12	2	4	-	
Axeheads, etc.	19	15	18	90	3+	21	11	49	-	

[1] also called microdenticulates in Europe.

HH (ME) = Hambledon Hill, Dorset, main enclosure; HH (Ste) = Hambledon Hill, Dorset, Stepleton enclosure; Et = Etton, Cambridgeshire; WH = Windmill Hill, Wiltshire; BH = Briar Hill Northanptonshire; Sta = Staines, Surrey; Ab = Abingdon, Oxfordshire; MC = Maiden Castle, Dorset.

Figure 5. Comparison of the total numbers of selected flint implement types from eight causewayed enclosures in England annotated to emphasise the functional puzzle of serrated flakes and the numerical domination of three categories (Incorporating numerical data from Saville 2002, p. 98, tab. 10.1).

Thus a key question which is integral to the functional puzzle is " why serrate? ". In modern steel tools serrated edges give a " grip and rip " advantage to *cutting* softer substances, for example serrated kitchen knives for vegetables such as tomatoes, bread knives for crumbly bread or cake, curved and toothed-edge sickles for hand reaping in parts of the Mediterranean, and lastly, the distinctive saw edge for cutting wood. However, it is not so easy to find modern examples of serrated edges used in a *transverse* motion. The most obvious might be to " shred " fibres or " grate " substances; in both cases it is what comes off that is the desired end product. However, there is one traditional " serrated " edge where the serrations are designed to *stop* the implement cutting too deeply into the surface being scraped as in this case it is the material left behind which is the desired end product. The implement (illustrated in Hurcombe 2007, fig. 7.4) is a bone tool used in hide-working. It has a sharp edge angle in profile but a rounded edge in plan view into which a series of small nicks are cut. As the tool wears down it is both re-sharpened and the notches are re-cut. The " serrations " help prevent the sharp edge digging into the hide surface too deeply. The solution to the functional puzzle of the serrated flakes needs to consider these kinds of issues and draw into the argument a reason why this distinctive edge shape is desirable.

PLANT FIBRES FOR CORDAGE AND TEXTILES: THE ARCHAEOLOGICAL EVIDENCE

Even where there is pictorial and written evidence for the use of fine cloth at a site and exceptionally good archaeological preservation, direct evidence of textile production is rare and archaeologists are instead required to think differently about some other kinds of evidence (e.g. Bronze Age Akrotiri

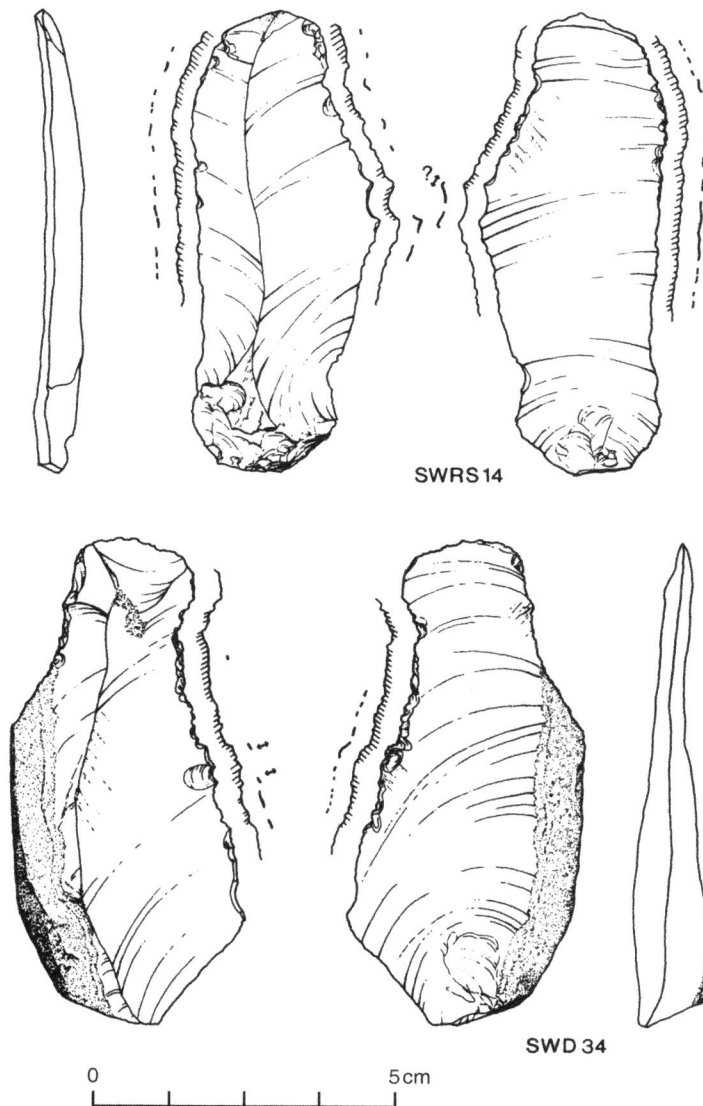

Figure 6. Other archaeological tools from the Neolithic Sweet Track flints from the Somerset Levels Project, UK, with less regular edge scarring whose function may be similar to the pieces recognised as " serrated tools ".

with textiles depicted on the painted frescoes, shells and plant species that may have been used for dye colours Tzachili 1990, Panagiotakopulu *et al.* 1997). Mindsets which include the possibilities for and knowledge of cordage and textiles production processes are as important as the extant archaeological evidence.

It is recognized that the survival of organic remains in the archaeological record is rare but it is also true that even within organic remains, there are particular biases against finding evidence for plant-based cordage, basketry and textiles, and the significance of small amounts of a wide range of evidence from pollen and plant remains is often overlooked (see Hurcombe 2000 for a fuller discussion). Animal fibres such as wool survive

better than vegetal fibres (Hald 1980, Barber 1991). Futhermore, the very act of fibre preparation (the stripping of extraneous materials and other plant cell material away from the fibres) ensures that only one plant structure survives on which to make any specific identification. That, coupled with degradation due to soil conditions and issues of conservation, makes it difficult to identify fibres to species level, and then only where good reference collections are available. Often the identification can only be listed as " bast fibres ", or " vegetal matter ". Such statements do not encourage investigation into a specific *chaîne opératoire* for plant and tree species.

Other categories of archaeological evidence such as pollen analysis could be expected to provide some evidence for plants whose fibres are suitable for cordage and textile production. Flax pollen is certainly a signal to consider fibre production, yet few attach true significance to the amounts recorded. Hall (1989) showed the poor traveling and survival qualities of flax pollen when the samples he collected from directly beneath flax plants showed only 4 % *Linum* pollen, while samples collected from only 1 metre away only occasionally showed *Linum* and then only as a single grain. Thus, for the archaeological record, the presence of a single pollen grain of flax could signify a " field " of flax within a metre of the sample or that the sampled deposit was strongly associated with the processing of flax. The issues for nettle pollen are similarly complex.

If nettle (*Urtica dioica*) pollen is present on a site it is most often seen as a " weed " species because of its propensity to grow in disturbed ground. However, *Urtica* pollen is thin-walled and thus fragile and has been shown to have a greatly reduced representation (Spieksma *et al.* 1994) where in airborn sampling *Urtica* pollen accounted for 30,6 % of the total, but only 2,1-0,7 % in samples collected on the ground (in this study from moss cushions). Thus although *Urtica* can be well-represented (Caseldine 1981, Proctor *et al.* 1996), it may also be under-represented and much will depend upon the context of the sample. Pits can provide especially rich samples and may also preserve seeds capsules and fibres as organic remains. For example, a flax-retting pit was identified on such evidence at the Bronze Age site of West Row, East Anglia (Martin and Murphy 1987). Similarly, retting pits with flax seeds and capsules have been found at another Bronze Age site, Reading Business Park and the pits also contained many *Urtica dioica* seeds which led the excavators to conclude that nettle fibre preparation should not be dismissed (Moore and Jennings 1992). Thus abundant evidence, identified to species, within a feature which can be seen as part of the process of fibre extraction, can be identified positively as evidence for flax retting but is more cautiously mentioned for nettle. If other methods of production were used in the Bronze Age or earlier, the lack of a feature to link the evidence would make fibre production even more difficult to substantiate in the archaeological record. For example retting can be achieved in rivers or ponds, or as dew retting in the fields: there would be no features to sample.

Similar issues affect the ability to see processing and weaving tools. Where weaving devices use inorganic weights, as in some upright looms, the evidence survives; but the backstrap loom tensioned by the weaver's body leaves no archaeological trace, and a five metre long ground loom tensioned against two posts sunk into the ground leaves two widely-spaced postholes which cannot be attributed to a specific purpose (Hurcombe 2000, 2007). Likewise the choice of tools for processing could affect the survival of evidence in the archaeological record. A wooden weaving sword or scutching knife is unlikely to survive, but bone versions of such tools might (e.g. Soffer 2004). The processes of fibre extraction have to be careful and to some extent gentle in order not to damage the desired end product; this leaves distinctive roles for stone tools with acute edges as primarily harvesting and cutting tools to gather the raw materials and make them into a format suitable for processing, and, perhaps to slit, split or trim off the material, while stone tools with high edge angles and blunt edges are primarily for scraping off extraneous material or to provide a blunt softening edge. The latter set of actions can also be performed by wood, bone and antler tools. The evidence from tool wear traces is thus likely to be in specific parts of the plant-processing *chaînes opératoires* and in order to correctly interpret the wear traces the variations and issues within the chain need to be understood as well as possible all the way from raw material to end product. A review of the known processes for producing baskets, fibres, cordage, mats and textiles has shown just how few of the processing activities need have been carried out with stone tools (see Hurcombe 1998 for a list).

Our ability to " see " such practices in the archaeological record is profoundly affected by prehistoric people: their cultural choices of raw materials, processes, tools, soil features and locations in the taskscapes and *chaînes opéra-toires* for plant-processing are at the same time both the information archaeologists desire, and conversely, the reason why some part or all of those processes remain unrecorded. However a short review of some of the archaeological evidence shows how much the mindsets and expectations of researchers as well as evidence bases affect discussions.

A summary of the evidence provided in Hurcombe (2000) bears reiteration with minor additions in the next few paragraphs. There are rich circum-alpine region finds and other areas of northern Europe with good organic preservation, as mentioned in other papers in this volume, but an awareness of the significance of nettles and tree bast for fibre production has not always been demonstrated. For example, Lundström-Baudais (1984) does explain the potential uses of the plant remains she identified at the Late Neolithic site of Clairvaux, Station III, giving *Urtica dioica* as a potential weaving plant, but Pals (1984, p. 316, tab. 1), investigating Aartswoud, a Late Neolithic coastal site in the Netherlands, lists the species identified from uncarbonised remains, indicating the relative abundance of *Urtica* in comparison to *Linum*, yet only draws attention to the latter. In addition, without individual reassessments, there are still likely to be mistaken identifications of flax at the expense of less well-known nettle or other bast fibres e.g. Henshall's report (Coles *et al.* 1964) on the cloth "plug" from Pyotdykes dated to the Late Bronze Age states that the material is flax, though another author suggests it may be nettle (Wild 1988). There is little discussion in Britain of tree bast sources as fabric fibres, but in continental Europe some sites have the majority of the yarns from such fibres (Médard 2000) and more bast fibres are being recognized (Reinhard 2000), and their importance emphasized (Rast-Eicher 2005). Furthermore, the field of textiles analysis has grown with more publications (e.g. Bender-Jørgensen 1992, Maik 2004, Pritchard and Wild 2005) and within this there is greater recognition of the tradition of making fabrics using twining and looping techniques rather than looms (Alfaro 1992, Rast-Eicher 1992, Rast-Eicher and Schweiz 1994).

Compared to the continent, the more recent British wetland sites have often not provided many organic artefacts. Though the preservation conditions in the Somerset Levels were suitable for the recovery of organic finds, the excavations focused on trackways which need not be the best context for the deposition of plant craftwork. However, the Somerset Levels Project demonstrated the use of moss and cotton-grass as a packaging material for some Bronze Age flakes, (Coles and Coles 1978, 1989, Brown 1986) and recovered a fragment described as "grass rope" consisting of grassy stems loosely twisted together (Coles *et al.* 1973, p. 288-289, pl. XXIX). However, one of the leaf-shaped arrowheads had the remains of nettle binding. The material was identified on the grounds that it was composed of plant fibre with apparent hair-bases on the surface similar to those on nettles (Coles *et al.* 1973, p. 291-292, fig. 18).

In contrast, the excavation of the wetland causewayed enclosure at Etton, Cambridgeshire (Pryor 1998) has a rich evidence base, probably because it was in at least some way a lived-in site. There is a fine hank of string fibres identified as *Linum usitatissimum* found in part of the causewayed enclosure ditch in a layer dated to the Middle Neolithic (phase 1A). It is 560 mm long and appears to be complete since there is no evidence that it has been cut. Small amounts of connective tissue were still present especially towards one end so it would seem to be unused (Taylor 1998, p. 157, pl. 174). The value of such finds is not as novel revelation but as rare confirmation of what we assume to be present in the organic material culture repertoire. Certainly, it is assumed that there are textiles in the British Neolithic, but the implicit assumption has been that these are of animal fibre. Henshall (1950) did comment that the main pastoral element of the Southern British Neolithic is cattle, with relatively small proportions of sheep. It is also known that the Neolithic sheep had less wool and more kemp which was a much coarser material (Barber 1991, Christiansen 2004, Ryder 1983, 2005). Etton also has an (unfinished?) antler comb (Armour-Chelu 1998, p. 286, fig. 244) and a range of flint tools, including serrated-edge flakes. There are 52 from Middle Neolithic contexts and 23 from Late Neolithic contexts. The parallels with the continental material in terms of both the edge detail and macroscopic polish traces suggest that these tools performed similar tasks. If the functions assumed for the continental material hold true, then at Etton, a dominant tool type from the Middle Neolithic contexts might tentatively be associated with the production of plant fibre. *Urtica dioica* is certainly well represented among the plant taxa, in addition to the find of flax string. Table 66 (Nye and Scaife 1998, p. 290-291) shows that for many of the enclosure ditch samples dated to the Middle Neolithic, and for the interior pit F505, *Urtica dioica* is a major component and many of the other species are potentially useful for cordage and basketry products. Some of the species listed are also recorded associated with flax growing according to Pals and van Dierendonck (1988, p. 245, tab. 3), though a similar growing environment might be expected to account for this. The local environment at Etton would certainly have included suitable materials for fibres, cordage and basketry and areas of water for retting or soaking processes. Some of the enclosure ditches are known to have been waterlogged and could have been suitable for retting, if their use as such was appropriate to the contemporary society. The simply made small wooden forks found in ditch segments 4 and 5 (Taylor 1998) would be useful tools for raising soaking bundles of plants or bark materials to check on their condition or even have functioned as distaffs (pers. comm. Anne Batard and Linda Mårtensson). Taylor (*op cit.*) suggests that the thick corky bark mostly found in ditch segment 5 may have been brought to the enclosure, possibly for tanning. It is worth noting that soaking plant fibres in tannic acid will aid in allowing resistant plant material to take up a dye and that leached woodash can also be useful in aiding retting or dyeing processes (Barber 1991). The heeled wooden points (Taylor 1998) may have been awls for basketry and textile work. I have argued previously (Hurcombe 2000) that taken altogether, much of the evidence from Etton could be interpreted as part of cordage, basketry and fabric production activities. To my knowledge, no other interpretations have run counter to this argument.

THE *CHAÎNES OPÉRATOIRES* SUGGESTED BY ETHNOGRAPHIC, HISTORICAL AND BUSHCRAFT SOURCES

It is clear that the processing of bast fibres and nettle fibres is not well understood compared to those fibre sources for which written records survive (e.g. Wild 1970). Season of harvest, age and species of plant and processing techniques are all important sources of variation. In some ways this situation mirrors the variation inherent in skin processing techniques. Three main sources can be used in combination to offer generic issues and specific examples. Archaeologists are used to employing historical and ethnographic sources, but bushcraft sources are also useful in understanding practical issues of materiality (Hurcombe 2007). The discussion below identifies a related set of retting-based processing techniques broadly comparable across flax, hemp and nettle as summarized in figure 7 and is followed by more specific discussion of nettles since this plant is far less well known. The sources used to inform this discussion are many and varied and some discuss several different kinds of materials and processes so the list cannot easily be split up except to say that nettles and tree bast fibres accounts are especially useful from Northwest coast of North America and bushcraft sources. Thus the bibliography contains many sources which are used to inform these general accounts rather than listed as reference here.

Flax fibres can require seven different operations to process them, making them the most difficult to prepare according to Wild (1970). Practices described across time and space (e.g. Baines 1989) follow common themes as figure 7 shows. There are tantalising hints at specific practices for flax and hemp growing and processing which might also relate to nettle. For example the following comparison uses information on flax and hemp processing from Wild (*op cit.*) who states that for flax and hemp the seeds are thickly sown to encourage the plants to grow tall and straight (nettles growing close together in ditches and rich damp environments likewise grow tall and straight); flax and hemp are harvested by pulling by hand (flax roots are long and straight making pulling easy; the root system of nettles and its propensity to propagate by runners makes hand pulling unlikely although this is possible where the roots are shallow or in soft soil); flax, and in some cases hemp too, is harvested before it is ripe i.e. before the woody core hardens but if harvested too early the flax fibres are soft and weak (nettles also have woody stems and may have the same issues of allowing the plant to attain maximum height and the fibres to mature whilst perhaps not allowing the core to harden, however much depends upon the processing techniques for extracting the fibres and there are accounts where the nettles are harvested in the Autumn or after first frosts); hemp fibres are uneven but strong and hemp is known for its qualities as a rope while, in contrast, the desirable qualities of flax are its tensile strength and durability, making it a good sailcloth (Bender-Jørgensen 2005, Cooke and Christiansen 2005, Möller-Wiering 2005), yet it has fine fibres which enable high quality fine fabrics to be made, it has a smooth handling quality and bleaches to a very light colour which can have colour symbolism for purity and cleanliness (nettle fibre is likewise, fine, strong, smooth and bleaches in the sun, although pond retted nettle is brown and dew retted nettle is grey and there are accounts of nettles being dyed with bark to render them invisible for fishing lines).

Nettles (*Urtica spp.*) are so named because of a common Indo-European word root indicating spin, sew, bind (Grieve 1931). They were especially preferred for fishing lines by American Northwest Coast societies because of their durability and toughness (Stewart 1977) and were used as wrappings for leisters and fish hooks at Mesolithic Tybrind Vig (Andersen 1987). The main ethnographic records of the processing and use of nettle are from Nepal (Dunsmore 1985), the Northwest Coast societies of America (Stewart 1973, 1977, Gustafson 1980) and North American Indians (Densmore 1929). A detailed discussion by Grieve (1931) indicates that it was also used in

	Flax	Hemp	Nettle
Harvest plant	Pull stalk	Pull stalk but the male plants ripen earlier than the female plants so a double harvest is necessary	Can pull plants up if soil is loose or roots are only shallow, but can also cut stems near the ground
	Leave to dry	Leave to dry	Leave to dry
Removal of seed heads	Seeds can be shaken off when dry or rippled with a wide toothed comb to remove the seeds	Seeds can be removed before the plant is harvested or combed out after harvest	Leaves and seeds drop off after a few days of drying, or can be removed by hand or with the aid of a tool
Preparation to loosen fibres via retting	A: By immersion in still or flowing water for 2-3 weeks B: By leaving dried stalks on the surface of fields into the autumn, known as dew retting	A or B	A or B possible or neither: see below
	Leave to dry	Leave to dry	
	Breaking the stalks and extracting the bundles of fibres and outer bark can be achieved by pounding them with wooden mallet or similar implement to break up the tougher core and loosen the outer bark and core from the bast fibres which are gathered together. A flax break can also be used whereby a stationary series of wooden slots and upstanding edges have a dovetailing complementary series of slots and edges set into an arm hinged onto the stationary set which can be raised and lowered with one hand as the stalks are drawn across with the other hand	The dried stalk can be pulled past a crooked finger several times which cracks the woody core into short sections all the way down the length of the plant. The core can then be stripped away and the outer bark removed all by hand. Wild (1970: 29-30) describes this as an evening activity undertaken by groups of women in rural Italy	Ray Mears describes lightly flattening the stem with a stone or other blunt implement to split it down its length. The fibres and outer bark can then be peeled away from the woody core with care
	Scutching allows the gathered fibres to be lightly beaten by a thin blunt edge, often a large wooden "knife", over another thin edge so that the remaining outerbark clinging to the fibres is broken up and beaten off	As flax	Could be as for flax
	Hackling occurs by drawing a handful of long fibres across a board with spaced spikes so that any remaining outer bark or inner core fragments are combed out of the fibres which remain neatly aligned	As flax	Could be as for flax

Figure 7. A comparison of the *chaînes opératoires* for flax, hemp and nettle (featuring retting).

Northern Europe until recently and, intriguingly, by Germany and Austria during the first world war to relieve the shortage of material for cloth. The nettle has also been the subject of more recent interest (Dreyer *et al.* 1996). The modern trials found that the fibre was more difficult to extract from nettles than flax, although processing was similar. Also, interestingly for this discussion of management, Grieve suggests that flax was an easier crop to grow. Nettles do grow wild and grow in many places, but for the finest fibres, the nettle had to grow in rich deep soil so that tall, very fibrous plants resulted. " The most valuable sort of nettle in regard to length and suppleness is most common in the bottom of ditches, among briars and in shaded valleys, where the soil is a strong loam. In such situations the plants will sometimes attain a great height, those growing in patches on a good soil, standing thick, averaging five to six feet in height, the stems thickly clothed in lint. Those growing in poorer soils and less favorable situations, with rough and woody stem and many lateral branches, run much to seed and are less useful, producing lint more coarse, harsh and thin " (Grieve 1931, p. 576).

The same source also states that the collection of nettles in Germany produced 2,7 million kg in 1916 without cultivation, but in time self-propagated plants were insufficient. This is not so surprising if the figure of 40 kg of the material to produce one shirt is accepted. The historical and ethnographic evidence suggests nettles would have been a useful plant for cordage and textile production and that the better stands of the plants near a settlement might have been reserved for producing finer items.

In contrast to Grieve, Mear's (2002, 2005) bushcraft experience leads him to value not the green but the red stemmed nettles for fibre production. I could find no botanical descriptions which stated a sub-species difference, but personal observation suggests that the red-stemmed nettles grow in full sun while the green ones are found in shadier conditions. Thus the colour of the stem is perhaps a factor of chlorophyll pigmentation according to growing conditions, but there does seem to be a genuine difference in the kind of fibres produced in these differing conditions which is clearly colour-signaled to anyone wanting to exploit specific qualities, and it is these differences which explain why Grieve's account stressing fine fibres for manufactured textiles from dried and spun fibres preferred one set of growing conditions for fineness of fibre, whilst Mear's who wanted an immediate, easily processed and tough " string ", favoured those nettles growing in other conditions.

For tree bast fibres two acts of separation and one of softening form the basis of most processing. The outer bark must be separated from the inner bark fibres if the finest materials are to be produced and if the outer material is thick, then this is even more the case. The wood and bast fibres must also be separated but this is usually done at the start of the process. Often, the ethnographic accounts show that the bark is peeled off living trees in long but thin strips so that the tree will not die. In these methods, outer and inner bark come off the wood together and subsequent processing can peel or beat the outer bark off the inner fibres. Depending upon the degree of fineness and softness the bark can be separated longitudinally or beaten, or in some cases soaked or retted so the bast fibres split into fine laminations.

Even this short discussion of nettle and other materials processing highlights the common features and the variation on which the experimental programme could draw. The experiments below tried various of these techniques but in the end used different methods to achieve the separations of outerbark, inner bark and wood for bast fibres as explained below.

USING SERRATED FLINT EDGES IN FIBRE PROCESSING *CHAÎNES OPÉRATOIRES*: THE RESULTING EXPERIMENTAL TRACES

Figure 8 outlines the experimental programme. The programme is informed by the ethnographic practices described but also there is a series of key issues arising directly from the archaeological wear traces which influenced the nature of the experimental programme and can be summarized as follows.

- Longevity. Use times for the archaeological tools must have been high, in the order of hours, days or even weeks of use at the task, so the experiments have a minimum period of an hour to establish a baseline for inter-experiment comparisons with much longer use times desired for more promising lines of experimentation.

- Lightness of action and contact material. There must be a reason for the lightness of action on the contact material, so it will be important to consider the end product as much as the process.

- Distinctive distribution of the most intensely-polished areas. The tip of the serrations should be the main point of contact and the material should be either narrow in itself or there should be a reason why a larger material has been split into narrower sections for working with the tool.

- The practical constraints of season, age and condition of plants. Plants should be worked in broadly appropriate seasons of the year and use materials of a suitable age and condition to obtain a desirable end product.

- High silica presence. Such intense polishes result from the presence of silica, but this could be from a variety of forms as silica-rich plants such as grasses, cereals, reeds and rushes, but could also result from clay, dust, ochre or other such substances.
Since the silica-rich polish is characteristic the source must be from either the contact material or another material which is regularly present in the process for a reason.

- Focus on the serrated edge as a specific edge type. The experiments need to show why serration is effective for the task and to investigate what this kind of edge is good for and what traces different actions might leave on such an edge.

SCRAPING	Less than 180 minutes	More than 180 minutes
Willow *(Salix)*	**#51** Epidermis from green fresh willow stems; 1 cm diameter (60 minutes) **#82** Epidermis from cut willow stems 1 week old; 2 cm diameter (60 minutes) **#83** Epidermis from silica-rich? Smooth and shiny willow stems; 1 cm diameter (100 minutes)	**#107 epidermis from Grey willow saplings; 3 - 4cm diameter (3 hours)**
Lime *(Tillia)*	**#52** Epidermis from fresh lime shoots; 1 cm diameter (60 minutes) **#81** epidermis from lime shoots, 1 week old; 1 cm diameter (60 minutes) **#112** epidermis from bast fibres *after* removal of entire bark from stem; 1 cm diameter (30 minutes)	**#109 Epidermis from green young shoots; 1 cm diameter (3 hours)**
Nettle *(Uritica dioica)*	**#47 fibres from stems boiled in a woodash solution (60 minutes)** **#48 fibres from stems retted in water in a clay pit (clay adhering), (60 minutes)** **#49 fibres from stems retted in water and woodash in a clay pit (less clay adhering), (60 minutes)** **# 113 epidermis from lightly pond-retted stems (in pond for 5 days)**	**#105 epidermis and fibres from stems (6 hours)** **#106 epidermis only from stems using exceptionally light action (6 hours)**
Other	*Cutting actions for comparisons between plain and serrated edges:* **#35 cutting nettle with an unserrated edge (60 minutes)** **#34 cutting nettles with a serrated edge (60 minutes)** **#108** *cutting* nettles to use in experiments (50 minutes) *Transverse actions:* **#45 strip nettle leaves (180 minutes)** **#111** scrape pith from *Scirpus lacustris* (10 minutes only because not effective edge shape for task) **#110** scrape pith from *Juncus* (50 minutes but not very effective edge shape for task) **#116** scrape flax stems to process (5 minutes only as edge shape not effective for task) # **000** tried scorching epidermis in fire prior to scraping but ineffective as charring was uneven # **000** tried smearing clay over epidermis prior to scraping but ineffective as coverage uneven	

Figure 8. Serrated flint tool experiments performed in Exeter and Lejre in 2006.
Key experiments whose wear traces are described further are in bold.

- Recording methods. Some of the features of wear which characterize the archaeological tools above are visible at different magnifications and so for good comparisons to be made photographs should run across magnifications where this is relevant.

The hypothesis was thus that serrated flint tools could have played a role in bast fibre production for fine fabric and cordage and that the experiments would focus on the *chaînes opératoires* of plant and tree raw materials.

Methods and Materials

The experiments aimed to explore the different ways of processing nettle and tree bast fibres using a serrated flint tool in order to create a reference set of wear against which to judge the archaeological wear traces. Exploratory experiments conducted in 2005 and spring 2006 suggested the experimental programme needed to:

- Use the tools for comparable periods and in some cases these should be for longer periods; some comparative experiments were conducted for 60 minute periods to establish baselines of wear traces but three hours minimum seemed realistic where the action was producing effective results.

- Test a wider range of possible methods of nettle fibre processing and conduct similar experiments with tree bast fibres.

- Provide sets of fibres from different kinds of processing for evaluation by spinners and weavers to ensure that the fibres produced were of value.

Flint tools

A set of serrated flint tools was prepared by Bruce Bradley in Exeter, using flint from one source at Beer, Devon so that the raw material and style of flint tool were held as constant as possible. All were handheld as there is no convincing indication that the archaeological tools were hafted.

Nettles

Stands of tall nettles (ca 5 foot or more high) for the experiments were harvested at Lejre and stored indoors to be used in the experiments as required. A person can harvest ca 20 - 30 nettle stems in 5 minutes, but only ca 6 - 30 stems can be processed per hour for fibre depending upon the technique used. The experiments involving retting pits of necessity were not conducted at Lejre but in the Exeter University Experimental area, using locally harvested nettle stands. Both red and green stems were used but these intra-species differences were not thought to be so great as between species or between different *chaînes opératoires*.

Lime

At Lejre young lime shoots from an avenue of pollarded trees were used by courtesy of Ledreborg Castle. At Exeter, basal re-growth shoots from a managed lime tree were used courtesy of Exeter University Buildings and Estates Department.

Willow

At Lejre, young " grey willow " shoots were used initially but in August the bast did not easily separate from the wood for these young shoots whereas thicker willow shoots (typically 3 - 4 cm diameter) proved more suitable for the bast production *chaîne opératoire*. At Exeter, wild willow from the University grounds was used as well as commercially available varieties from the willow beds of Linda Lemieux, a local basketmaker.

Other

Scirpus lacustris, *Juncus* and flax were all available at Lejre and in Exeter and were used in exploratory experiments.

The sets of fibres from all the different processing methods were kept separate and those suitable fine fibres passed to spinners and weavers for evaluation. The tools were taken back to Exeter University for full documentation of the wear traces and comparison with archaeological traces.

Results

Twenty-three variations were explored which resulted in twenty-one experiments with stone tools suitable for wear analysis (numbered) as outlined in figure 8. These either allowed certain processes or additives to be ruled out or showed promise as possible matches when compared to the archaeological wear traces. The most important of these are given in bold and it is these whose wear traces are more fully described below and in the figures. Three sets of raw materials were investigated as willow, lime tree and nettle. The set of experiments described in figure 8 under " other ", documents the attempts to use serrated tools for fibre extraction of *Scirpus lacustris*, *Juncus* and flax that did not prove effective.

In addition, a series of directly comparable plant harvesting experiments exactly mirroring time, species and circumstance, were conducted using both plain and serrated edges in cutting actions. Figures 9 and 10 show the results for cutting nettles but such experiments were repeated with other plant species. The experiments demonstrate

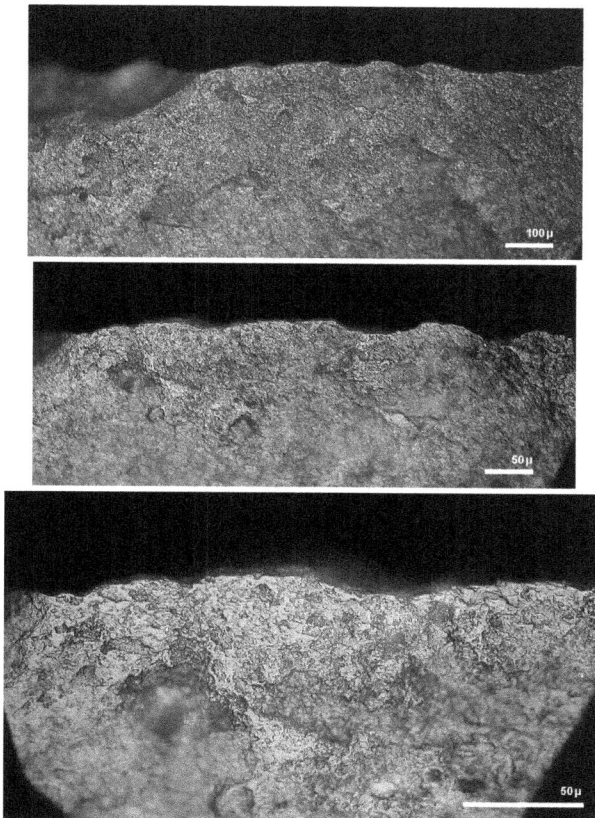

Figure 9. #35 cutting nettle with a plain flint edge
for 1 hour. OM 100X, 200X, 500X.

two issues. Firstly, if the serrated tools were used for comparatively short periods of time for harvesting plants the fairly light wear traces would be " overwritten " by the scraping actions which dominate the genre and secondly, the serrated tools made very effective plant harvesting tools. At first the plain edges cut better and the serrated edge simply cut well, but after 20 - 30 minutes the plain edges started to feel slightly dull although these were effective for the full 60 minutes. In comparison, the serrated edges simply maintained their good performance. These tools were especially effective when cutting fibrous plants or water-rich plants. The serrated tools were even effective at cutting the woody nettle stems. The wear traces may reflect a different contact action of " grip and rip " for the serrated tools, as across all the mirrored harvesting experiments undertaken, the serrated tools generally showed slightly less well developed wear traces than their plain-edged counterparts, although traces could build up more unevenly on edge prominences.

Using ideas taken from a range of ethnographic and other sources (e.g. Stewart 1973, 1984, Hurley 1979, Gustafson 1980, Kuoni 1981, Dunsmore

Figure 10. #34 cutting nettle with a serrated flint edge
for 1 hour. OM 100X, 200X, 500X.

1985, Baines 1989, Turner 1998, Mears 1990, 2002, 2005, Pole and Doyle 2004, Duer and Turner 2005), some investigative experiments with bast fibre processing, especially different forms of nettle processing were conducted prior to coming to Lejre. These involved scraping fibres off stems boiled in a solution of wood ash, or retted in pits. The resulting wear traces showed the polish *distribution* from scraping stems to be a close match to the archaeological tools, but not quite the right texture. The experiments from both stripping leaves and shredding previously removed plant fibre did not give the right wear distribution. In both cases the wear formed was not at its most intense at the tips, but was, in contrast, more intense in the recessed parts of the edge. This can be clearly seen in figure 11 showing the wear from #45 used in stripping the nettle leaves off the stem. A forceful action was required and small edge damage scars in the recesses can also be seen. Thus I am convinced from this range of experiments that the archaeological tools were used to scrape something broadly similar in size to a nettle stem or small sapling. However, with any task it is the end product which would have been the purpose of the tool. The experiments all resulted in fibres, but these were in different states: some were very

tangled together or contained a lot of extraneous ash/clay fragments which could not easily be removed. The experiments suggested other ways of processing using hearths and singeing the stems in wood ash, or smearing the stems with clay prior to further processing. Some archaeological fabric fragments have been identified as *tree* bast fibres so these bast fibre sources were also incorporated into the planned experiments.

Tree bast fibres

The *chaîne opératoire* for tree bast fibre requires the bast fibres to be separated from the outer bark, or epidermis, and also from the woody interior. For some purposes such as basketry it may not be necessary to remove the outer bark but in any finer basketry or fabric style applications the fibres would be more flexible if this extraneous material were removed. However the order in which the outer bark and the woody interior are separated from the fibres could differ as explained above. The pilot experiments suggested that it was much easier to remove the outer bark whilst the bast was still on-the-wood because this gave a hard surface and the material being scraped was held firmly in place. If the peeled complete bark was scraped to remove the outer layer the task was considerably more awkward (#112). Thus the common method of using the serrated tool was to scrape the outerbark off the bast fibres (fig. 12) and then make two careful slits lengthwise to enable the bast fibre layer to be cleanly peeled from the wood in two strips (fig. 13). This *chaîne opératoire* worked for both lime and willow shoots of ca 1 cm diameter which gave fine fibres during the growing season as the sap layer eases separation,

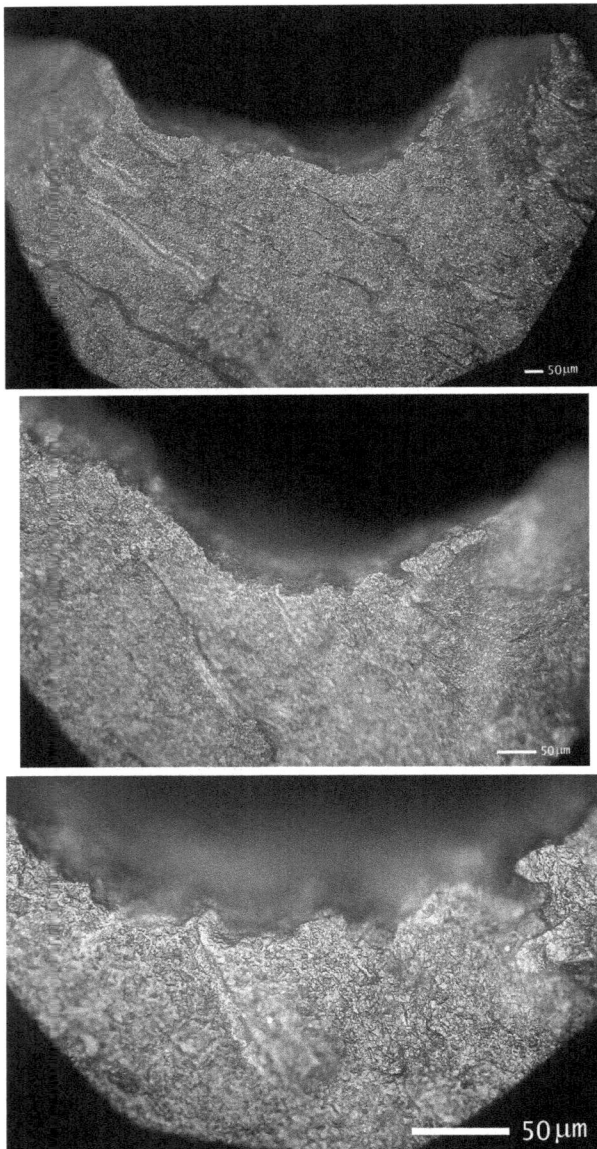

Figure 11. #45 stripping nettle leaves from the stems for 3 hours. OM 100X, 200X, 500X.

Figure 12. Scraping the epidermis off grey willow branches as the first stage of bast fibre preparation.

56

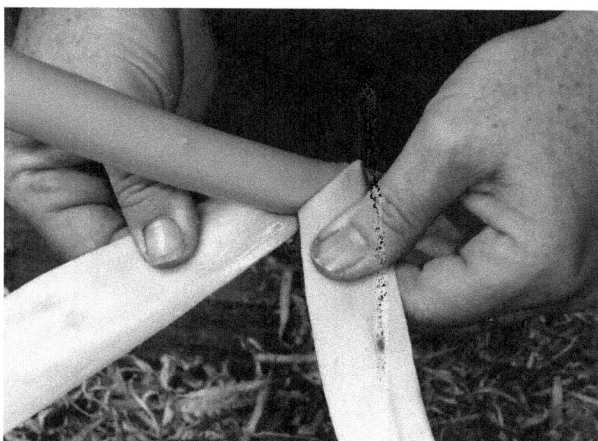

Figure 13. Peeling the bast fibres from the willow wood after carefully scoring them.

but it was found that in late August, thicker willow stems peeled better. The extracted fibres were boiled overnight in woodash and water; further experimentation with this technique could be useful in the future. Some variations in these experiments investigated very smooth shiny

Figure 14. #83 scraping epidermis from very shiny (silica-rich?) willow shoots for 100 minutes. OM 100X, 200X.

willow outer bark thought to be silica-rich (#83, fig. 14), but no silica-rich polish resulted after one hour of use and there was little difference between the wear traces for this type of willow, or between freshly cut 1 cm (#51) or 2 cm (#82) willow stems. Likewise there were little wear trace variations in scraping the outer bark from fresh lime shoots (#52) versus week old lime shoots (#81). Since the archaeological polish suggested long use of the tools, two experiments scraped willow (#107) and lime (#109) outer bark from bast fibres for 3 hours.

Figure 15 shows the resulting wear traces for lime. The microscopic edge scarring is broadly comparable but the polish was more widely distributed, and not so intense. The latter features may yet be affected by using the tools for longer so that areas of polish develop better linkage. Figure 16 shows comparable wear traces for grey willow clearly showing some edge-tip scarring but a brighter polish with better linkage. This shows some promise as a match for the archaeological traces although there are more striations in the experimental traces.

Nettle bast fibres

The accepted *chaîne opératoire* for flax fibre extractions is to ret (partially rot) the material in dew (with thanks to Anne Batard for describing this process) or in retting pits or slow moving water. Several investigative experiments tried to mirror these retting strategies for nettles but with some variations likely to introduce extra silica into the contact. To this end comparable 1 hour experiments were performed resulting in the wear traces shown in figure 17.

Tool #47 scraped fibres from the stems of nettles boiled in a solution of wood ash and water. The resulting fibres slid easily from the stems such that a tool seemed unnecessary and moreover the fibres were tangled and epidermis was still adhering to the fibres and acting as glue. Drying and pounding might have helped but it was also appreciated thanks to Anne Batard that these fibres were "overcooked" and very brittle. Whilst using a weaker solution of wood ash might improve some aspects the fibres would still be both tangled and stuck together with epidermis. Thus the *chaîne opératoire* here was ineffective. The wear traces were also weak.

Tool #48 scraped fibres off pit-retted nettles. There was a lot of clay adhering to the nettle

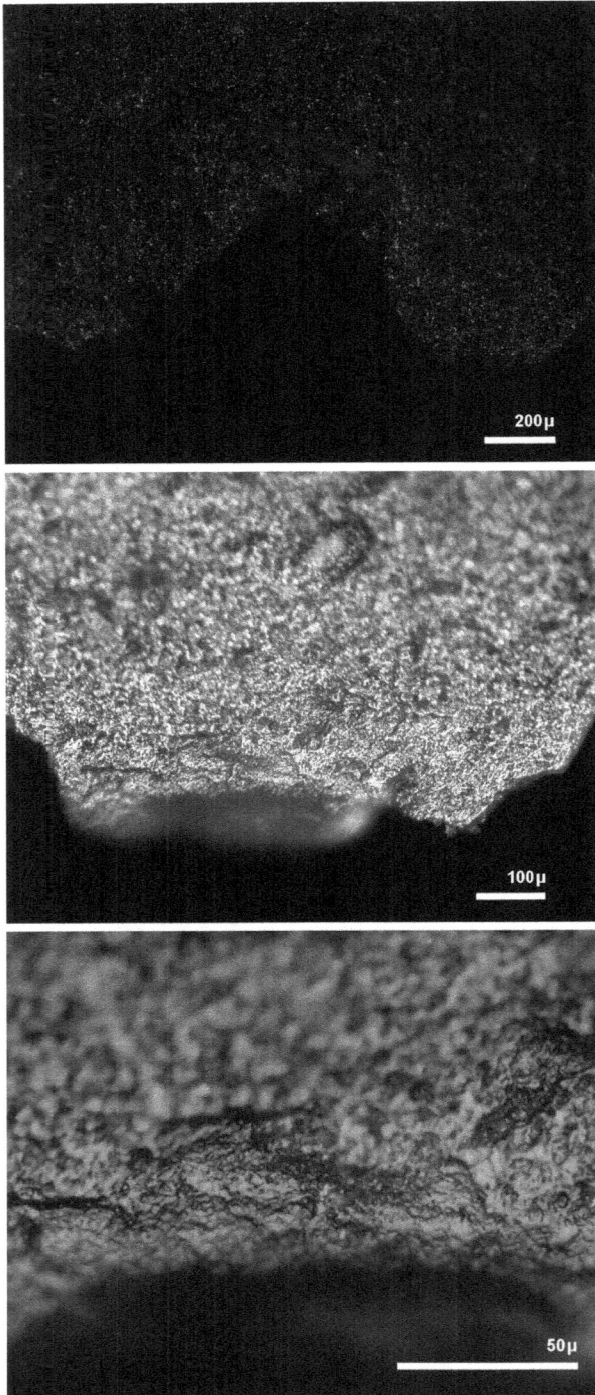

Figure 15. #109 scraping epidermis from lime shoots for 3 hours. OM 50X, 100X, 500X.

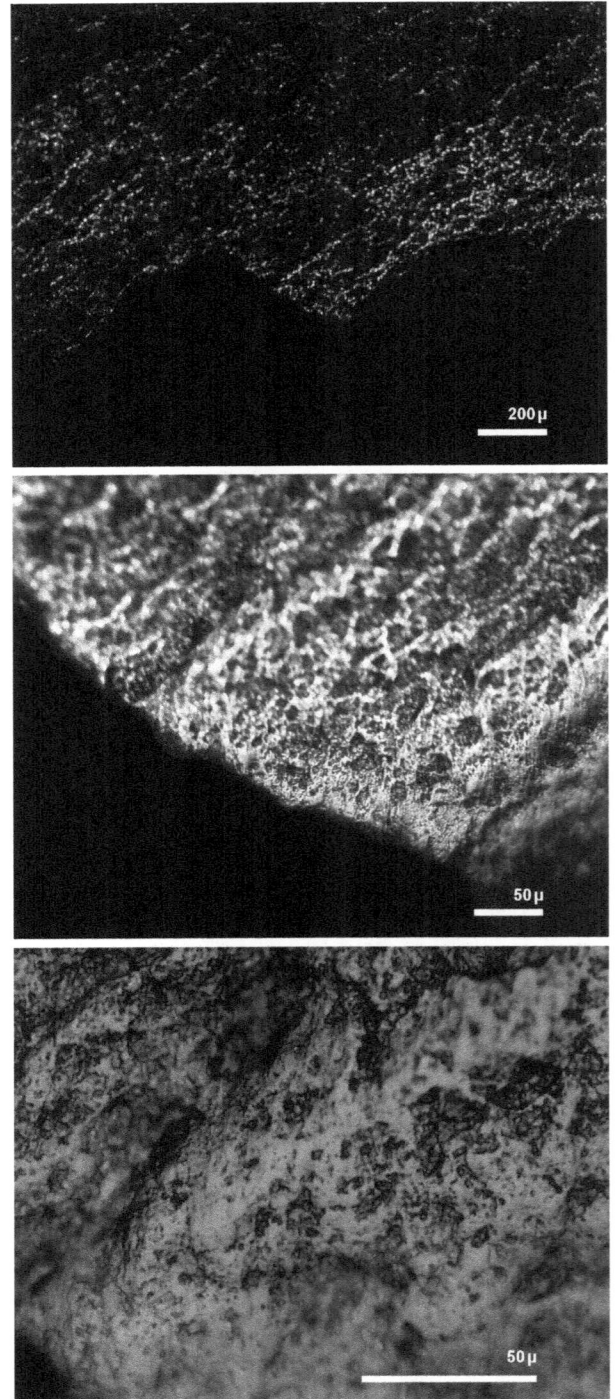

Figure 16. #107 scraping epidermis from grey willow for 3 hours. OM 50X, 200X, 500X.

stems and a strong stench. Again the fibres came away easily but epidermis was still present and the whole was a mass of clay, fibre and some epidermis which although in some long strands was difficult to clean and deemed an ineffective *chaîne opératoire*. The wear traces were bright and rounded with many striations caused by the presence of the clay particles.

Tool #49 scraped fibres from nettle stems retted in similar pits for exactly the same amount of time but with wood ash added. The caustic woodash seems to have acted to disperse the clay as the stems pulled from the pit were cleaner with less clay adhering than those in experiment #48. The stench was not so marked and the fibres still separated easily but again the fibres extracted although in some long hanks included epidermis although with less clay and would require cleaning and aligning.

The *chaîne opératoire* was not especially effective and the wear traces still had many fine striations compared with the archaeological wear traces.

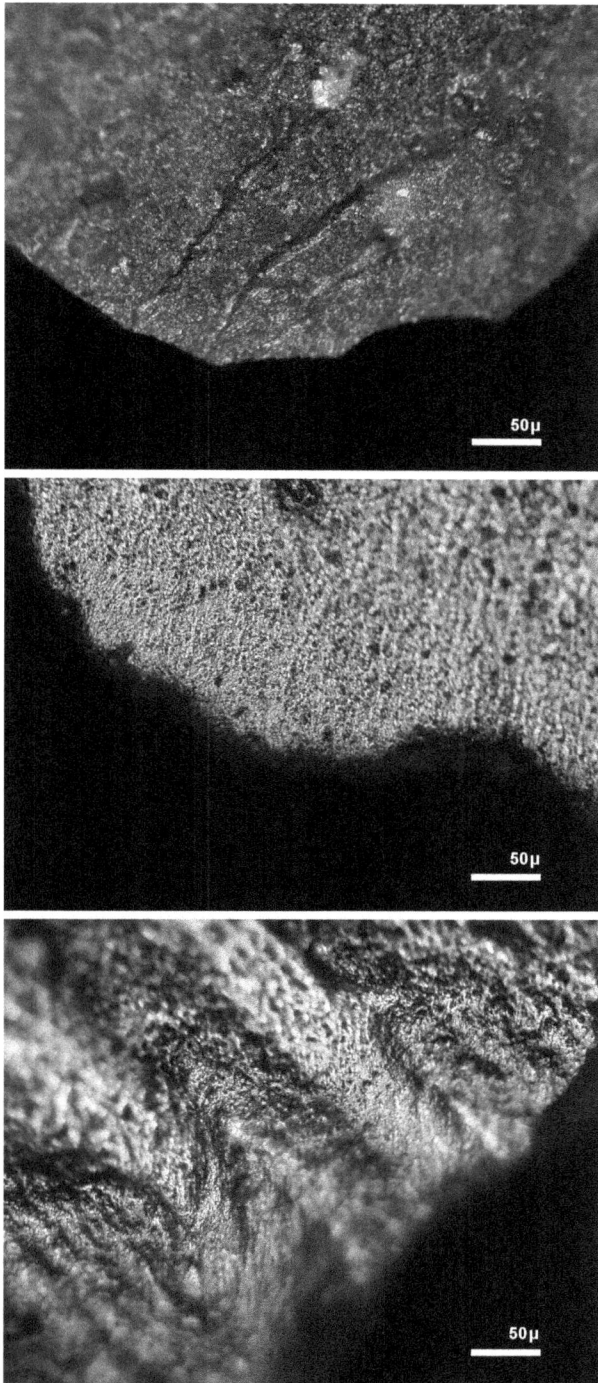

Figure 17. Wear traces from scraping for 1 hour nettle stems boiled in a woodash solution (#47), nettle stems retted in a clay pit (#48) and nettle stems retted in a clay pit with woodash (#49). All OM 200X.

A different *chaîne opératoire* was tried out by using the serrated tool #105 to scrape both epidermis and bast fibres from the stems of nettles. This was extremely effective in removing all the fibres. However the resulting fibres were tangled and still had epidermis adhering, as seen in figure 18. This material was excellent as insulating " shoe-hay " and would also have been effective as tinder or absorbent padding but as a material for spinning up it was more problematic. Anne Batard showed that it was possible to sort the tangled fibres by hand (hedgehog and other combs were also tried) and then spin them successfully using a spinning hook (fig. 19). By conducting the fibre extraction and the spinning of the fibres side by side in this way it was obvious that there would be trade-offs in the *chaîne*

Figure 18. Scraping nettle stems to remove both the epidermis *and* fibres from the woody core efficiently, creating a bundle of unaligned fibres which Anne Batard is straightening. Nettle fibres stripped by hand from split stems are drying in the background.

Figure 19. Anne Batard spinning freshly-stripped nettle fibres with a spinning hook after sorting the fibres by hand.

opératoire for fine cordage production because in all likelihood in prehistory it would have been the same person performing all the tasks. There seemed little point in saving time in fibre stripping if there were high wastage rates in combing the fibres or much greater input of time in sorting the fibres by hand to untangle them. Also Anne Batard clarified that longer hanks of fibres were both easier to spin and resulted in stronger yarn. Thus there were some problems with the effectiveness of the *chaîne opératoire*. The wear traces however showed a promising match with the archaeological ones (fig. 20). The polish was concentrated on the tips of the serrations and was bright with a similar distribution over the surface high and low spots. Its texture was not as smooth but this could be developed further by greater longevity even though the tool was used for six hours.

A further experiment essentially used the familiar *chaîne opératoire* for tree bast fibre extraction on nettle stems: the tool was used to scrape the epidermis from the fibres on the nettle stems in an

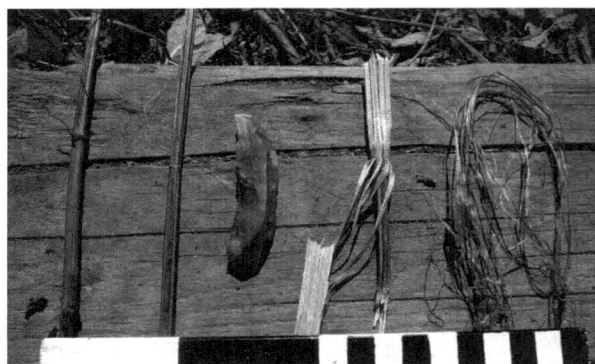

Figure 21. A new *chaîne opératoire* for nettle processing. Nettle stems were stripped of their leaves (left), then the outermost layer was lightly scraped off, (second from left), the stem was gently flattened to split it in half longitudinally and then the fibres were gently peeled away from the woody core (second from right) to leave the fibres aligned and in long hanks (right).

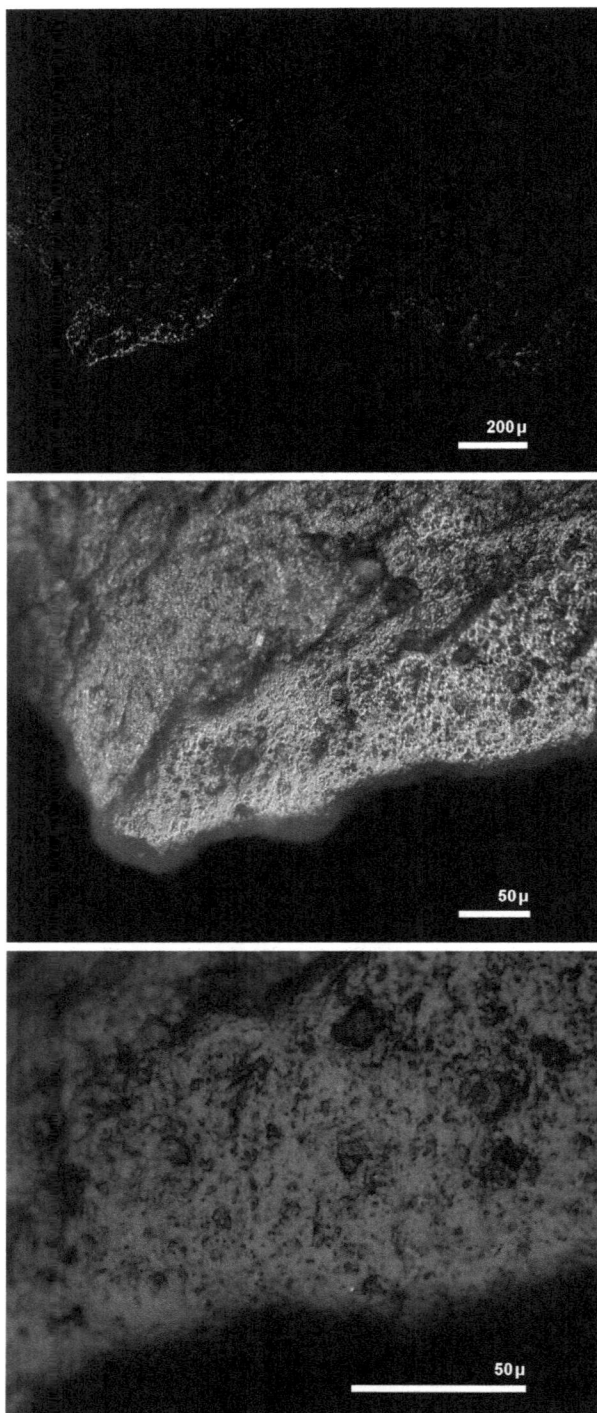

Figure 20. #105 Scraping epidermis *and* fibres from nettles for 6 hours. OM 50X, 200X, 500X.

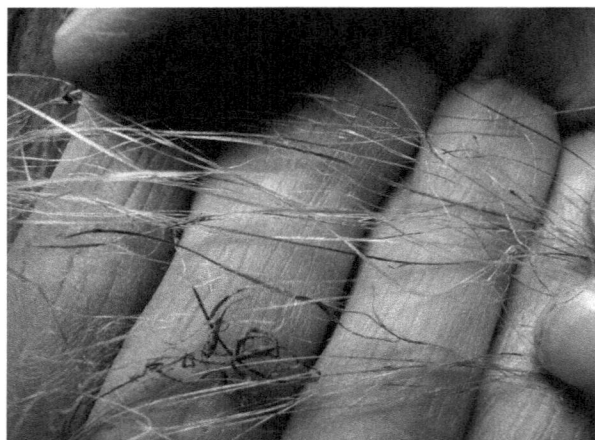

Figure 22. The nettle fibres produced in this way were fine and fairly clean although some epidermis remained.

exceptionally light and careful action. Stems with only the epidermis removed could be split by using a pebble or wood billet to lightly flatten the stem causing a split to open. The split was widened carefully so that the stem parted longitudinally. These split stems were bent apart in the centre and the fibre gradually peeled away from the wood. Figure 21 shows the resulting

long length of fibre with a small remnant of the wood remaining to be peeled off. Although some epidermis remains much of the fibre is clean (fig. 22). This technique arose from a general knowledge of the ethnographic and bushcraft *chaînes opératoires* coupled with direct interaction with the experienced weaver at Lejre and show the benefits of close collaboration and using the tool for a task in a *chaîne opératoire* rather than to " make wear ".

The technique was perfected in the last day or so of the project and it remains for Anne Batard to fully assess the resulting fibres. Overall, the processing was slow but the fibres were presented in a condition suitable for spinning strong and fine yarn. The resulting wear traces on the tool (fig. 23) were the best match of all for the archaeological wear traces although, once again, the polish might be a better match if the tool were used for longer: six hours of experimental wear shows a promising match but is still an order of magnitude away from the development of the wear on the archaeological tool. The wear is concentrated on the tips of the serrations and at higher magnifications shows linkage and some intensity. It is thought that the stinging hairs on the nettle which have a silica skeleton provide a source of silica to aid the intensity of the polish (fig. 24) and some phytolith residue extractions are planned to elucidate this.

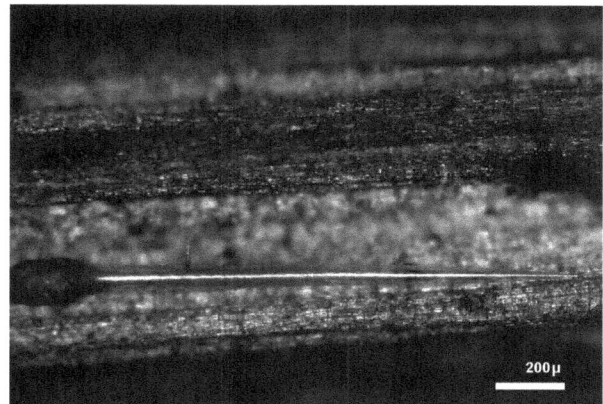

Figure 24. The surface of a nettle stem showing the silica structures of the stinging hairs. OM 50X.

CONCLUSIONS

The key aspects of the project were experiments to investigate the function of serrated (saw edged) flint tools from the earlier Neolithic of Britain and Denmark, and the processing of a very common

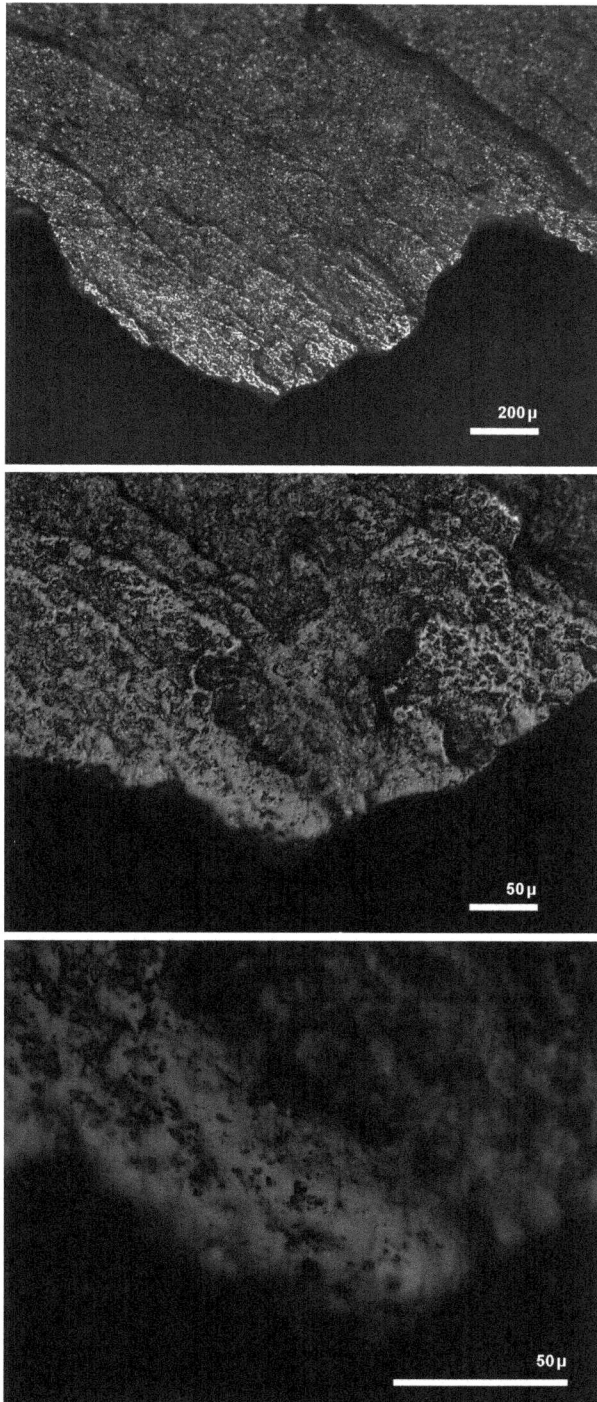

Figure 23. #106 Scraping epidermis *only* from nettle stems for 6 hours. OM 50X, 200X, 500X.

plant, stinging nettles (*Urtica dioica*), to extract fibres for cordage and fabric. The experiments harvested and processed nettles by separating the fibres from the woody stem and removing the external epidermis to leave very fine fibres. Methods included soaking the nettles in a pond, also in pits, boiling the nettle with woodash (to make a caustic solution) and hand extractions with no tools. The weaver's opinion of some of the fibres produced, and the ease with which they could be spun up into yarn, suggested invaluable further refinements of some of the tool actions and methods used. Tools were also used to process tree bast fibres of willow *(Salix)* and lime *(Tilia)*. These fibres were also processed by a range of methods using the flint tools and different solutions. Taken altogether the experiments covered a good range of different plant fibres and the use of the flint tools for different stages in the production process. They resulted in a better understanding of the performance characteristics provided by this very distinctive serrated edge. In addition, the tools were all used for substantial periods of time (three hours for more promising tasks) giving better sets of wear on the tool edges for comparison under the microscope with the wear traces observed on the archaeological tools. In the past, the production of cordage for all manner of purposes, and the manufacture of finer yarns for making clothing, would both have been time-consuming and very important cultural tasks.

The serrated flint tools are amongst the most prolific finds on earlier Neolithic sites and the experiments performed at Exeter and Lejre offer further understanding of the role these tools performed in the production of items of organic material culture for which so few other traces survive. The most promising matches to the archaeological traces are for processing willow and nettle fibres, with nettle the closest. Further experiments this season are expected to clarify and strengthen these conclusions.

The experimental wear traces have suggested novel aspects of the plant processing *chaînes opératoires* for which there is no strong ethnographic data, but where practical experiments have demonstrated the efficacy of the processes and provided better matches to the archaeological record than exist when following known ethnographic *chaînes opératoires*. In particular, the hypothesised function addresses some of the key issues identified. It explains why a serrated edge is useful: it does not dig too deeply into the fibres as the desired product but shallowly scrapes

off the unwanted outerbark layer. It shows why the most intense wear is on the tips of the serrations and why the zone of most intense polish can be laterally restricted to *circa* 1 cm; if the contact between stem and tool is always in the same place, and if it is the smaller stems and saplings which are being worked as whole round materials with fibres still on the wood, then the contact zone is in the same place and of limited lateral extension. It suggests a source for the silica which gives the silica-rich polish characteristics: nettle stems have silica hairs which form part of the contact material if the outer stem is the key contact surface. Conversely the remaining issues of poorer aspects of the match can be logically explained. The polish is less well developed; if the proposed function is correct the archaeological tools may well have seen days or weeks of use. Some of the archaeological pieces have variations in smoothness and striations; these could be due to season, age, growing conditions and the presence of extra factors such as dust or silt.

If the closest traces identification is correct, then nettle processing may have been one of a related set of important cordage and textile activities. Furthermore it would seem that nettles followed the tree bast fibre suite of *chaînes opératoires,* in scraping outer bark from fibres before removing them from a stem. Such evidence for organic crafts is rare, but perhaps the combination of archaeological, ethnographic and experimental data says something yet more interesting. Manual processing gives fine fibres through considerable time investment and care, revealing both a different attitude to time and the level of material culture engagement which made the elaboration of the production of material culture items worthwhile. Alternatively, the evidence could suggest that there were deeply conservative traditions within society which kept the same processing techniques for fine fabrics of nettle as had traditionally been used in tree bast fibre production.

Thus archaeological strands of evidence and very specific information from wear traces, general ethnographic, historical and bushcraft accounts, tempered by practical experiments, have all contributed to the suggestion of a novel *chaîne opératoire* in an undervalued craft sphere as a potential solution to a widespread functional puzzle in time and space. The growth in popularity of serrated-edge flakes and then later decline could signal a shift in the materials and or processes used in fine cordage and fabric *chaînes opératoires*.

ACKNOWLEDGEMENTS

The "organics from inorganics" project was initially funded by a Leverhulme pilot project grant which allowed me to test out some of the issues and especially focused on cordage and basketry evidence from flints and pottery. I should like to thank the following: Valérie Beugnier and Phillipe Crombé for inviting me to participate in the international seminar in Ghent which proved such a productive interchange of research ideas; Lejre experimental research centre for the grant to conduct experiments at their centre, and especially to Anne Batard one of their resident experts who shared her knowledge of spinning and weaving with me; Exeter University grounds staff for permissions to use various ponds, pits and plants; my colleague Bruce Bradley for producing flint blades; and, last but not least, Mandy Pike, Emily Pike and Amaranta Pasquini, all of whom acted as assistants for conducting and recording the experiments.

AUTHOR'S ADRESS

Linda HURCOMBE
University of Exeter
Department of Archaeology
EXETER EX4 4QE - UK
L.M.Hurcombe@exeter.ac.uk

REFERENCES

ADOVASIO, J. M., SOFFER, O. and KLÍMA, B., 1996. Upper Palaeolithic fibre technology: interlaced woven finds from Pavlov I, Czech Republic, c. 26,000 years ago. *Antiquity,* 70, 526-534.

ALFARO, C., 1992. Two Copper Age tunics from Lorca, Murcia (Spain). *In:* L. Bender-Jørgensen and E. Munksgaard, eds. *Archaeological textiles in Northern Europe: report from the 4th NESAT symposium, 1-5 May 1990 in Copenhagen.* Tidens Tand Nr. 5. Copenhagen: Konservatorskolen det Kongelige Danske Kunstakademi, 20-30.

ANDERSEN, S., 1987. Tybrind Vig: A submerged Ertebølle settlement in Denmark. *In:* J. Coles and A. Lawson, eds. *European Wetlands in Prehistory.* Oxford: Clarendon, 253-280.

ARMOUR-CHELU, M., 1998. The animal bone. *In:* F. Pryor, ed. *Etton: Excavations at a Neolithic causewayed enclosure near Maxey, Cambridgeshire 1982-7.* London: English Heritage, 273-287.

BAINES, P., 1989. *Linen: Hand Spinning and Weaving.* London: Batsford.

BAMFORD, H., 1985. *Briar Hill.* Northampton: Northants Development Corporation.

BARBER, E.J.W., 1991. *Prehistoric Textiles.* Oxford: Princeton University Press.

BENDER-JØRGENSEN, L., 1992. *North European textiles until AD 1000.* Aarhus: Aarhus University Press.

BENDER-JØRGENSEN, L., 1994. Ancient costumes reconstructed: a new field of research. *In:* G. Jaacks and K. Tidow, eds. *Archäologische Textilfunde - Archaeological textiles, Textilsymposium Neumünster, 4-7 May 1993, NESAT V.* Neumünster: Textilmuseum Neumünster, 109-113.

BENDER-JØRGENSEN, L., 2005. Textiles of seafaring: an introduction to an interdisciplinary research project. *In:* F. Pritchard and J. P. Wild, eds. *Northern archaeological textiles, NESAT VII textile symposium in Edinburgh, 5-7 May 1999.* Oxford: Oxbow, 65-69.

BOCQUET, A., 1994. Charavines il y a 5000 ans. *Les Dossiers de l'Archéologie,* 199, 1-104.

BRADSHAW, R. H. W., COXON, P., GREIG, J. R. A. and HALL, A. R., 1981. New fossil evidence for the past cultivation and processing of hemp (*Cannabis sativa* L.) in Eastern England. *New Phytologist,* 89, 503-510.

BROWN, A., 1986. Flint and chert small finds from the Somerset levels. Part 1: The Brue Valley. *Somerset Levels Papers,* 12, 12-27.

CASE, H. and WHITTLE, A., 1982. *Settlement Patterns in the Oxford Region: Excavations at the Abingdon Causewayed Enclosure and Other Sites.* London: Council for British Archaeology.

CASELDINE, C., 1981. Surface pollen studies across Bankhead Moss, Fife, Scotland. *Journal of Biogeography,* 8, 7-25.

CHRISTIANSEN, C. A., 2004. A reanalysis of fleece evolution studies. *In:* J. Maik, ed. *Priceless invention of humanity – textiles, NESAT VIII, 8-10 May, 2002.* Acta archaeologica Lodziensia Nr. 50/1. Łódź: Łódzkie Towarzystwo Naukowe, Instytut Archeologii i Etnologii PAN, 11-17.

COLES, J. M. and COLES, B. J., 1978. Bronze Age Implements from Skinner's Wood. *Somerset Levels Papers,* 4, 114-121.

COLES, J. M., COUTTS, H. and RYDER, M. L., 1964. A Late Bronze Age find from Pyotdykes, Angus, Scotland, with associated gold, cloth, leather, and wood remains. *Proceedings of the Prehistoric Society,* 30, 186-98.

COLES, J., HIBBERT, F. A. and ORME, B.J. 1973. Prehistoric roads and tracks in Somerset: 3 the Sweet Track. *Proceedings of the Prehistoric Society,* 39, 256-293.

COOKE, B. and CHRISTIANSEN, C., 2005. What makes a Viking sail? *In:* F. Pritchard and J.P. Wild, eds. *Northern archaeological textiles, NESAT VI: textile symposium in Edinburgh, 5-7 May 1999.* Oxford: Oxbow, 70-74.

DENSMORE, F., 1929. *Chippewa Customs.* Washington: Smithsonian.

DEUR, D. and TURNER, N. J., eds, 2005. *Keeping it living: traditions of plant use and cultivation on the Northwest coast of North America.* Washington: University of Washington Press and Vancouver: University of British Columbia Press.

DREYER, J., Drayling, G. and Feldmann, F., 1996. Cultivation of stinging nettle *Urtica dioica* (L) with high fibre content as a raw material for the production of fibre and cellulose: qualitative and quantitative differentiation of ancient clones. *Journal of Applied Botany,* 70, 28-39.

DUNSMORE, S., 1985. *The Nettle in Nepal: A Cottage Industry.* Surbiton: Land Resource Development Centre.

GODWIN, H., 1967. The Ancient Cultivation of Hemp. *Antiquity,* 41, 42-49, 137-138.

GRIEVE, M., 1931. *A Modern Herbal.* London: Jonathan Cape.

GUSTAFSON, P., 1980. *Salish weaving.* Vancouver: Douglas & McIntyre.

HALD, M., 1980. *Ancient Danish Textiles from Bogs and Burials.* Copenhagen: National Museum of Denmark.

HALL, V., 1989. The historical and palynological evidence for flax in County Down. *Journal of Ulster Archaeology,* 52, 5-9.

HENSHALL, A., 1950. Textiles and weaving appliances from prehistoric Britain. *Proceedings of the Prehistoric Society,* 16, 130-162.

HURCOMBE, L. M., 1998. Plant-working and craft activities as a potential source of microwear variation. *Helinium,* 34 (2), 201-209.

HURCOMBE, L. M., 2000. Time, skill and craft specialisation as gender relations. *In:* M. Donald and L. Hurcombe, eds. *Gender and Material Culture in Archaeological Perspective.* London: Macmillan, 88-109.

HURCOMBE, L., 2007. *Archaeological artefacts as material culture.* London: Routledge.

HURCOMBE, L., in press. Looking for prehistoric basketry and cordage using inorganic remains: the evidence from stone tools. *In:* L. Longo and N. Skakun, eds. *Prehistoric technology 40 years later.* BAR International Series. Oxford: Archeopress.

HURLEY, W. M., 1979. *Prehistoric Cordage: Identification of Impressions on Pottery.* Washington: Aldine Manuals on Archaeology, 3.

JUEL JENSEN, H., 1994. *Flint tools and Plant Working.* Aarhus: Aarhus University Press.

KOOISTRA, L., 2006. Fabrics of fibres and strips of bark. *In:* L. P. Louwe Kooijmans and P. F. B. Jongste, eds. *Schipluiden: a Neolithic settlement on the Dutch North Sea coast c.3500 CAL BC.* Analecta Praehistorica Leidensia, 37/38. Leiden: Faculty of Archaeology, Leiden University, 253-259.

KUONI, B., 1981. *Cestería Tradicional Ibérica.* Barcelona : Ediciones del Serbal.

LUNDSTRÖM-BAUDAIS, K., 1984. Palaeoethnobotanical investigation of plant remains from a Neolithic lakeshore site in France: Clairvaux, Station III. *In:* W. van Zeist and W. A. Casparie, eds. *Plants and Ancient Man: Studies in Palaeoethnobotany.* Rotterdam: Balkema, 293-305.

MAIK, J., ed., 2004. *Priceless invention of humanity - textiles: NESAT VIII, 8-10 May, 2002.* Acta Archaeologica Lodziensia Nr. 50/1. Łódź: Łódzkie Towarzystwo Naukowe, Instytut Archeologii i Etnologii PAN.

MARTIN, E. and Murphy, P., 1987/1988. West Row Fen, Suffolk: A Bronze Age fen-edge settlement site. *Antiquity,* 62, 353-358.

MEARS, R., 1990. *The Survival Handbook.* Oxford: Oxford Illustrated Press.

MEARS, R., 2002. *Bushcraft.* London: Hodder and Stoughton.

MEARS, R., 2005. Making string from nettle, skills section, *Bushcraft* [DVD]. London: BBC.

MÉDARD, F., 2000. *L'artisanat textile au Néolithique : l'exemple de Delley-Portalban II (Suisse) 3272-2462 avant J.-C.* Préhistoire 4. Montagnac : Monique Mergoil.

MERCER, R., 1981. *Excavations at Carn Brea, Illogan, Cornwall, 1970-3.* Truro: Cornwall Archaeological Society.

MIDDLETON, H. R., 1998. Flint and chert artefacts. *In:* F. Pryor, ed. *Etton: Excavations at a Neolithic causewayed enclosure near Maxey, Cambridgeshire 1982-7.* London: English Heritage, 215-256.

MÖLLER-WIERING, S., 2005. Textiles for transport. *In:* F. Pritchard and J. P. Wild, eds. *Northern archaeological textiles, NESAT VII textile symposium in Edinburgh, 5-7 May 1999.* Oxford: Oxbow, 75-79.

MOORE, J. and JENNINGS, D., 1992. *Reading Business Park: A Bronze Age Landscape.* Oxford: Oxford Archaeological Unit.

MOOREY, P. R. S., 1982. A Neolithic ring ditch and Iron Age enclosure at Newnham Murren, near Wallingford. *In:* H. Case and A. Whittle, eds. *Settlement Patterns in the Oxford Region: Excavations at the Abingdon Causewayed*

Enclosure and Other Sites. London: Council for British Archaeology.

NYE, S. and SCAIFE, R., 1998. Plant macrofossil remains. *In*: F. Pryor, ed. *Etton: excavations at a Neolithic causewayed enclosure near Maxey, Cambridgeshire, 1982-7.* London: English Heritage, 289-300.

PALS, J. P., 1984. Plant remains from Aartswoud, a Neolithic settlement in coastal areas. *In*: W. van Zeist and W.A. Casparie, eds. *Plants and Ancient Man: Studies in Palaeoethnobotany.* Rotterdam: Balkema, 293-305.

PALS, J. P. and VAN DIERENDONCK, M. C., 1988. Between flax and fibre: cultivation and processing of flax in a mediaeval peat reclamation settlement near Midwoud (Prov. Noord Holland). *Journal of Archaeological Science*, 15, 237-251.

PANAGIOTAKOPULU, E., BUCKLAND, P. C., DAY, P. M., DOUMAS, C., SARPAKI, A. and SKIDMORE, P., 1997. A lepidopterous cocoon from Thera and evidence for silk in the Aegean Bronze Age. *Antiquity,* 71, 420-29.

POLE, L. and DOYLE, S., 2004. *Second Skin.* Exeter: Royal Albert Memorial Museum and Art Gallery.

PRITCHARD, F. and WILD, J. P., eds, 2005. *Northern archaeological textiles, NESAT VII textile symposium in Edinburgh, 5-7 May 1999.* Oxford: Oxbow.

PROCTOR, M., YEO, P. and LACK, A., 1996. *The Natural History of Pollination.* London: Harper Collins.

PRYOR, F., ed., 1998. *Etton: Excavations at a Neolithic causewayed enclosure near Maxey, Cambridgeshire 1982-7.* London: English Heritage.

RAST-EICHER, A., 1992. Neolitische Textilien im Raum Zürich. *In:* L. Bender-Jørgensen and E. Munksgaar, eds. *Archaeological textiles in Northern Europe, report from the 4ᵗʰ NESAT symposium, 1-5 May 1990 in Copenhagen.* Tidens Tand Nr. 5. Copenhagen: Konservatorskolen det Kongelige Danske Kunstakademi, 9-19.

RAST-EICHER, A., 2005. Bast before wool: the first textiles. *In:* P. Bichler, K. Grömer, R. Hofmann-De Keijzer, A. Kern and H. Reschreiter, eds. *Hallstatt textiles: technical analysis, scientific investigation and experiment on Iron Age textiles.* BAR International Series 1351. Oxford Archaeopress, 117-131.

RAST-EICHER, A. and SCHWEIZ, E., 1994. Gewebe im Neolithikum. *In:* G. Jaacks and K. Tidow, eds. *Archäologische Textilfunde – Archaeological textiles, Textilsymposium Neumünster, 4-7 May

1993, NESAT V.* Neumünster: Textilmuseum Neumünster, 18-26.

REINHARD, J., 2000. Textiles et vannerie. *In :* D. Ramseyer, ed. *Muntelier/Fischergässli : un habitat au bord du Lac de Morat (3895 à 3820 avant J.-C.).* Archéologie fribourgeoise, 15. Fribourg : Editions Universitaires, 200-205.

RYDER, M. L., 1983. *Sheep and Man.* London: Duckworth.

RYDER, M. L., 1999. Probable fibres from hemp in Bronze Age Scotland. *In*: G. Jones, ed. *Environmental Archaeology 4.* Oxford: Oxbow, 93-95.

RYDER, M. L., 2005. The human development of different fleece-types in sheep and its association with the development of textile crafts. *In:* F. Pritchard and J.P. Wild, eds. *Northern archaeological textiles, NESAT VII textile symposium in Edinburgh, 5-7 May 1999.* Oxford: Oxbow, 122-128.

SAVILLE, A., 1990. *Hazleton North, Gloucestershire, 1979-82.* London: Historic Buildings and Monuments Commission.

SAVILLE, A., 2002. Lithic artifacts from Neolithic causewayed enclosures: character and meaning. *In*: G. Varndell and P. Topping, eds. *Enclosures in Neolithic Europe.* Oxford: Oxbow, 91-105.

SOFFER, O., 2004. Recovering perishable technologies through use wear on tools: preliminary evidence for Upper Palaeolithic weaving and net making. *Current Anthropology*, 45, 407-413.

SOFFER, O, ADOVASIO, J. M. and HYLAND, D. C., 2000. The " Venus " figurines: textiles, basketry, gender, and status in the Upper Palaeolithic. *Current Anthropology,* 41, 511-537.

SPIEKSMA, F. T. M., NIKKELS, B. H. and BOTTEMA, S., 1994. The relationship between recent pollen deposition and airborne pollen concentration. *Review of Palaeobotany and Palynology*, 82, 141-145.

STEWART, H., 1984. *Cedar - tree of life to the Northwest Coast Indians.* Vancouver: Douglas and Macintyre.

STEWART, H., 1973. *Indian Artefacts of the Northwest Coast.* Seattle: University of Washington Press.

STEWART, H., 1977. *Indian Fishing on the Northwest Coast.* Vancouver: Douglas and McIntyre.

TAYLOR, M., 1998. Wood and bark from the enclosure ditch. *In*: F. Pryor, ed. *Etton: Excavations at a Neolithic causewayed enclosure near Maxey, Cambridgeshire 1982-7.* London: English Heritage, 115-160.

TURNER, N. J., 1998. *Plant technology of first

peoples in British Columbia. Vancouver: University of British Columbia Press.

TZACHILI, I., 1990. All important yet illusive: looking for evidence of cloth-making at Akrotiri. *In* D.A. Hardy, ed. *Thera and the Aegean World 3*. London: Thera Foundation, 327-349.

VAN GIJN, A., 1989. *The Wear and Tear of Flint: Principles of Functional Analysis Applied to Dutch Neolithic Assemblages*. Analecta Praehistoria Liedensia, 22. Leiden: Faculty of Archaeology, Leiden University.

VAN GIJN, A. 1998. Craft activities in the Dutch Neolithic: a lithic viewpoint. *In:* M. Edmonds and C. Richards, eds. *Understanding the Neolithic of North-Western Europe*. Glasgow: Cruithne, 328-350.

VAN GIJN, A., 2006. Implements of bone and amber: a Mesolithic tradition continued. *In*: L. P. Louwe Kooijmans and P. F. B. Jongste, eds. *Schipluiden: a Neolithic settlement on the Dutch North Sea coast c.3500 CAL BC*. Analecta Praehistorica Leidensia, 37/38. Leiden : Faculty of Archaeology, Leiden University, 207-224.

WILD, J. P., 1970. *Textile manufacture in the Northern Roman provinces*. London: Cambridge University Press.

Wild, J. P., 1988. Textiles in Archaeology. Aylesbury: Shire Archaeology.

ACQUISITION ET TRAITEMENT DES MATIÈRES TEXTILES D'ORIGINE VÉGÉTALE EN PRÉHISTOIRE : L'EXEMPLE DU LIN

Emmanuelle MARTIAL et Fabienne MÉDARD

Résumé : En Europe occidentale, la culture du lin est attestée dès le VIe millénaire, mais il faut attendre le IVe millénaire pour qu'apparaissent les premiers témoignages de son utilisation textile. Les procédés d'acquisition et de transformation de la fibre de lin pendant la préhistoire étant difficiles à aborder sur la seule base des vestiges matériels, il est nécessaire de diversifier les sources d'informations pour mieux comprendre les modes opératoires mis en œuvre : à ce titre, les données paléo-environnementales, ethno-historiques et expérimentales constituent des aides précieuses. En effet, l'artisanat textile lié aux périodes très anciennes ne peut être appréhendé que par le biais d'une approche pluri-disciplinaire, aussi complète que possible. Il s'agit d'une démarche novatrice qui consiste à suivre le déroulement d'une chaîne opératoire et à rechercher, pour chacune des étapes, les témoins archéologiques correspondant. Ce travail ne peut être effectué que dans le cadre d'une véritable collaboration entre plusieurs spécialistes et archéologues, depuis les opérations de terrain jusqu'à l'analyse des contextes et du mobilier.

Mots-clés : lin, textiles, préhistoire, expérimentation, approche archéologique et ethno-historique.

Abstract: In western Europe flax was cultivated from the 6th millennium, but the first evidence for its use in making textiles dates only to the 4th millennium. As the acquisition and transformation of flax fibre in prehistoric times are aspects that are difficult to examine on the sole basis of material remains, one must rely on diverse sources of information to understand the processes involved. Palaeoenvironmental, ethno-historical and experimental data are particularly helpful in this respect. Indeed the question of textile manufacture in very early times can only be addressed through a multi-disciplinary approach, as broad as possible. This approach should focus on the different phases of the « chaîne opératoire », trying to find corresponding archaeological evidence. This demands a close collaboration between specialists of different disciplines, both during field work and in the post-excavation phase (analyses of finds and contexts).

Keywords : flax, textiles, Prehistory, experimentation, archaeological and ethno-historical approach.

En Europe occidentale, les vestiges archéologiques prouvent que le travail des matières textiles est maîtrisé dès le Néolithique et, bien que les témoignages fassent défaut pour les périodes antérieures, il est probable que les connaissances dans ce domaine soient plus anciennes.

Si les restes textiles témoignent de savoir-faire techniques incontestables, les données liées aux processus de transformation des matières premières sont quasiment inexistantes. Le mode d'extraction de la matière textile au Néolithique est mal connu : aucun outil ne peut y être associé de façon certaine. Il s'agit pourtant d'une étape essentielle, car le traitement infligé à la matière première influe sur la qualité du produit fini.

Nous avons choisi de fonder notre propos sur le processus d'acquisition des fibres de lin et cela pour plusieurs raisons :

- Il s'agit tout d'abord d'un matériau emblématique de l'artisanat textile.

- Ensuite, son utilisation est attestée dans de nombreuses régions, aux époques les plus anciennes. En Europe occidentale, les premières traces d'agriculture du lin datent du VIe millénaire et les premiers vestiges textiles, du Ve millénaire (Médard 2006a). Depuis lors, son usage n'a jamais cessé et, bien qu'il soit aujourd'hui travaillé industriellement, le lin participe encore abondamment à la confection textile. Sans doute cette pérennité a-t-elle permis de conserver et de transmettre les savoir-faire liés aux processus de transformation de cette fibre de tige. Dès l'époque romaine, Pline l'Ancien (1964 - réédition) en exposait le détail ; 2000 ans plus tard, son témoignage montre que les méthodes ont très peu varié jusqu'à nos jours.

- Enfin, en exposant les modes d'acquisition et de traitement des fibres de lin, nous souhaitons verser quelques éléments à la connaissance de disciplines complémentaires à l'archéologie, telles que la tracéologie et l'archéobotanique. Le recoupement de données issues de plusieurs horizons pourrait éclairer certains aspects d'un processus opératoire dont le détail nous échappe encore pour les périodes préhistoriques.

Pour aborder cet aspect de l'artisanat, nous disposons de plusieurs sources d'informations : la culture matérielle, les structures archéologiques mises en évidence sur le terrain, les données paléobotaniques, les analyses tracéologiques, l'expérimentation et la documentation ethno-historique. Les données archéologiques qui pourront, à titre d'exemple, illustrer notre propos proviennent de différents horizons géographiques, liés en particulier à nos contextes d'étude respectifs.

Le schéma opératoire lié à l'acquisition des fibres de tige est à peu près semblable pour toutes les plantes qui en fournissent : le lin, l'ortie, le chanvre, le jute … Traditionnellement, l'extraction et la préparation des fibres se divisent en sept temps : la récolte, l'égrenage, le rouissage, le séchage, le battage, le teillage et le peignage.

ACQUISITION DE LA MATIÈRE PREMIÈRE

La récolte

De nos jours, et depuis plusieurs siècles, le lin est traité en plante annuelle ; semé au printemps (mars-avril), il est récolté au début de l'été. Entre la levée et la maturité, une centaine de jours s'écoulent au cours desquels la tige atteint sa hauteur maximale entre 80 cm et 120 cm.

Le moment choisi pour récolter la plante a une incidence sur la qualité des fibres. En Europe, au début du XXᵉ siècle, la récolte du lin à différents stades de maturation est encore pratiquée (Bonnétat 1919). Une récolte précoce fournit une filasse douce, fine, facile à blanchir, mais peu abondante. Elle prive également l'agriculteur des semences ; ce dernier doit alors disposer de pieds suffisamment nombreux pour qu'une partie d'entre eux parviennent à maturité et assure la pérennité de la culture. À l'inverse, une récolte tardive permet d'obtenir des graines de qualité, au détriment de la filasse, robuste et dure. Généralement, pour ménager les semences et la fibre, la récolte a lieu au cours du mois de juillet. Les documents divergent peu quant aux méthodes traditionnellement utilisées pour récolter le lin : les tiges sont arrachées manuellement avec leurs racines, pour laisser le terrain propre et obtenir des fibres longues (fig. 1).

On ne saurait dire si tel était le cas au Néolithique ; peut-être le lin était-il laissé en terre plusieurs saisons d'affilées. Or, la durée de végétation a également une incidence sur la qualité de la filasse (Bonnétat *op. cit.*). En tant que plante vivace, le lin gagne en vigueur au fil du temps ; il est récolté par fauchage pour que la souche reste en terre et reprenne l'année suivante. Cette option, si elle évite les semailles répétées, nécessite un entretien du terrain et une protection assidue contre les animaux sauvages et domestiques. Traité ainsi, il pousse lentement, mais s'avère plus productif au terme de quelques années.

Dans ce domaine, l'examen répété des fibres archéologiques permettra sans doute d'obtenir des précisions. À ce titre, la matière expérimentale montre que les fibres des lins récoltés verts présentent des caractéristiques différentes de celles des lins parvenus à maturité (Médard 2005).

Figure 1. Arrachage du lin au début du XXᵉ siècle (nord de la France ?) (carte postale ancienne).

Les analyses palynologiques, carpologiques et anthracologiques menées à partir d'échantillons de sédiments recueillis lors de la fouille de sites néolithiques (nord de la France) permettent également de préciser les connaissances liées à la qualité des matières textiles utilisées en préhistoire : elles livrent notamment des informations sur les espèces végétales, sauvages et cultivées exploitées par les premières sociétés agro-pastorales dans le cadre de l'artisanat textile.

L'étude des phytolithes, microrestes siliceux d'origine végétale que l'on retrouve dans les niveaux archéologiques associés aux habitats, permet, pour sa part d'identifier certaines composantes paléobotaniques du paysage, en particulier les plantes utilisées sur les sites mêmes. On sait que les végétaux endémiques employés pour la sparterie, la vannerie ou le textile sont potentiellement variés et souvent présents aux abords des sites d'habitat. Cette situation prévalait encore, en Europe du nord-ouest, aux périodes historiques. Ainsi, les saules (*Salix* sp.*) et des espèces aussi diverses que la clématite (*Clematis vitalba* L.), la ronce commune (*Rubus fructicosus* sp.), le cornouiller sanguin (*Cornus mas* L.), le noisetier (*Corylus avellana* L.), le troène commun (*Ligustrum vulgare* L.), la viorne lantane (*Viburnum lantana* L.), le roseau commun (*Phragmites australis*), le jonc des tonneliers (*Scirpus lacustris* L.), le jonc épars (*Juncus effusus* L.), la massette (*Typha latifolia* L.), la grande ortie (*Urtica dioica* L.), le merisier (*Prunus avium* L.) et sans doute bien d'autres encore, étaient utilisés il y a à peine quelques décennies (Lieutaghi 1998, Bonnier nd, Caspar *et al.* 2005).

Sur le site néolithique final d'Houplin-Ancoisne « Rue M. Dormoy » (Nord, France), la présence de liber de tilleul a été détectée par l'analyse anthracologique. Mais c'est surtout le lin (*Linum usitatissimum* L.) qui y est abondamment représenté sous la forme de graines carbonisées parmi les macrorestes végétaux et, dans une moindre mesure, sous la forme de pollen (Martial et Praud sous presse). Plus récemment, la présence d'une semence de lin, dont la culture était jusqu'alors inédite au Néolithique ancien dans le nord de la France, a été découverte sur le site Villeneuve-Saint-Germain de Meaurecourt, dans les Yvelines (Durand *et al.* 2006). Si l'exploitation du lin pour les fibres textiles qu'il fournit plutôt que pour les vertus oléagineuses de ses graines n'est pas directement prouvée à travers les études carpologique et palynologique, c'est l'association des restes de cette plante avec des indices archéologiques indiscutablement liés au dérou-

lement d'activités textiles (fusaïoles et pesons) qui en fournit les arguments tangibles.

L'égrenage

À l'issue de la récolte, les tiges de lin sont liées en bottes puis engrangées en attente d'être égrenées. Cette opération consiste à placer les sommités des tiges de lin sur les dents d'un peigne, généralement en bois – l'égrugeoir – et à tirer énergiquement vers soi. Les capsules, bloquées par les dents du peigne, tombent à terre (fig. 2). Elles peuvent alors être exposées au soleil afin que, sous l'effet de la chaleur, elles s'ouvrent et libèrent les graines (Saudinos 1942). Une autre méthode consiste à procéder par battage ou à récolter les graines à la main sur le lin en pied. Cette dernière solution, testée en expérimentation, s'avère d'une relative efficacité en l'absence d'outillage spécifique (Rast-Eicher und Thijsse 2001). Les plus belles graines servent à la semence, les autres sont utilisées pour la fabrication de l'huile et l'alimentation animale (Ewers 1989).

Figure 2. Égrenage des tiges de lin à l'aide d'un égrugeoir solidement fixé sur une planche. Artisanat traditionnel (Suisse) (cliché : F. Médard).

Sur le plan archéologique, quelques outils pourraient avoir servi à égrener le lin. Les peignes à côtes sont évoqués pour cet usage et il est vrai que les tests expérimentaux se sont avérés relativement efficaces (Reinhard 2000). D'autres outils, plus ou moins énigmatiques, ont également pu être utilisés dans ce cadre : des « râteaux » en bois dont on ignore s'ils étaient ou non bloqués à la base, pourraient avoir servi à l'égrenage (Médard sous presse) (fig. 3). Il ne s'agit toutefois que d'hypothèses inspirées par la morphologie de ces ustensiles.

fibres de leur ciment interstitiel. La fin du rouissage se reconnaît au manque d'adhérence des fibres au bois de la tige. C'est l'opération la plus importante et la plus délicate dans la préparation des fibres de lin (fig. 4).

Plus ou moins poussé, le rouissage agit en trois temps sur la matière première : il désolidarise les faisceaux de fibres du reste de la tige, il désolidarise les faisceaux de fibres entre eux, et enfin, il divise les faisceaux de fibres. Un rouissage

Figure 3. Branches écorcées, appointées et ligaturées, formant un peigne.
Site néolithique de Pfäffikon-Burg (ZH, Suisse) (Les lacustres, 150 objets racontent 150 histoires 2004).

Le rouissage

Par un processus de putréfaction contrôlée, le rouissage consiste à dégrader et à éliminer les ciments pectiques qui environnent les faisceaux de fibres. Il s'agit d'une dégradation biologique due à l'action des bactéries et des moisissures. La dissolution des pectines entraîne la dissociation des

correctement mené dure assez longtemps pour diviser les faisceaux de fibres sans isoler complètement les fibres élémentaires. En conséquence, une filasse suffisamment solide pour subir les étapes ultérieures de la transformation textile sera rouie de sorte à conserver une partie des matières pectiques qui cimentent les faisceaux de fibres (Médard sous presse).

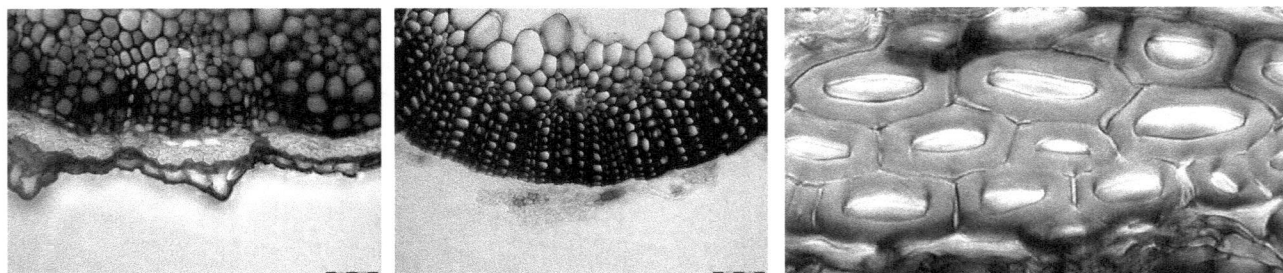

Figure 4. À gauche : tige de lin non rouie. On distingue, de bas en haut, l'épiderme, les groupes de fibres, le xylème (X100). Au milieu : tige de lin rouie. L'épiderme a totalement disparu et la plupart des faisceaux de fibres se sont détachés du xylème (X100). À droite : groupe de fibres (X400) (clichés : F. Médard).

On sait que le rouissage des lins verts est beaucoup plus rapide que celui des lins mûrs (Duhamel du Monceau 1762, Vétillart 1876). Ce phénomène s'explique par l'anatomie de la plante. Les fibres de lin vert ont des parois incomplètes (dépôt cellulosique inachevé) et sont peu comprimées : les faisceaux de fibres sont donc plus ou moins isolés à l'état naturel. L'action du rouissage, destinée à scinder les faisceaux de fibres, doit être rapide et stoppée avant d'atteindre les fibres élémentaires (Médard 2004).

Le rouissage peut être effectué de deux façons : par immersion, en eaux courantes ou stagnantes, ou à l'air libre. Par immersion, les bottes de lin sont maintenues sous l'eau entre 8 et 10 jours, mais l'opération peut durer jusqu'à 3 ou 4 semaines au printemps et seulement 4 à 6 jours en été. Roui à l'air libre, le lin est déposé à terre : l'action conjuguée de l'humidité, de la rosée, de la pluie, du vent et du soleil décompose la pectine des tiges. La durée du rouissage dépend des conditions climatiques. Cette méthode nécessite de nombreuses manipulations (le lin doit être retourné avant que la putréfaction ne gagne la filasse) et expose les tiges aux vers et aux rongeurs.

Les indices archéologiques liés au rouissage sont difficilement repérables, d'autant que cette étape ne nécessite pas forcément l'utilisation d'aménagements ou d'outillages spécifiques. Des concentrations de macrorestes végétaux pourraient éventuellement attester l'emplacement d'aires de rouissage, mais c'est rarement le cas. En revanche, les fouilles archéologiques révèlent parfois des aménagements particuliers. Par exemple, de grandes fosses caractérisées par leur profil en forme de fente figurent en contexte danubien dans l'est de la France, comme à Holtzheim (Bas-Rhin) (Lefranc et Arbogast 2000) et correspondent aux *Schlitzgruben* décrites par les archéologues allemands. Récemment, quatre fosses à profil en forme de fente ont également été mises au jour sur le site d'habitat Villeneuve-Saint-Germain fouillé à Meaurecourt « Croix de Choisy », sur la rive droite de l'Oise (Yvelines, France) (Durand *et al.* 2006, Dietsch-Sellami *et al.* 2006). De dimensions variables, elles ont une forme allongée et un comblement formé d'une alternance de couches de sédiments organiques et de couches de sédiments remaniés du substrat limoneux. L'étude carpologique et l'analyse des phytolithes incitent les auteurs à proposer une interprétation fonctionnelle de ces structures comme fosses destinées au rouissage de plantes textiles. Dans le comblement de ces fosses, la

présence de l'ortie est en effet associée à des indicateurs d'humidité.

Sur le site néolithique final de Houplin-Ancoisne « Rue M. Dormoy » (Nord, France), localisé en bordure du fond de vallée marécageux de la Deûle, une très grande fosse mesurant 12,4 m de long sur 3,2 m de large et 1,6 m de profondeur maximale (soit un volume d'environ 55 m³) a été creusée dans la partie basse du site actuellement soumise aux battements de la nappe phréatique (fig. 5).

Figure 5. Grande fosse de rouissage (?) du site néolithique final d'Houplin-Ancoisne « Rue M. Dormoy » (Nord, France) (cliché : I. Praud).

Constituée de deux alvéoles situées dans le prolongement l'une de l'autre, elle présentait une alternance de couches compactes d'ossements animaux, tapissant le fond et les parois, et de remblais de sédiments argileux présentant des recreusements très nets et comprenant de fins lits organiques. La stratigraphie, au sein de laquelle douze épisodes successifs ont été distingués, se terminait par une phase d'utilisation secondaire comme dépotoir avant d'être scellée par une ultime couche d'abandon. La culture attestée du lin, la présence de fusaïoles et de pesons en terre cuite suggéraient des activités textiles sur le site. Les données micromorphologiques et pédologiques confirment que la fonction de cette fosse est liée en particulier à la présence de matières organiques, aux phénomènes de stagnation d'eau et à la forte acidification du sédiment. La convergence d'éléments obtenus par une approche pluridisciplinaire la plus complète possible (paléo-environnementale, pédologique, archéologique) permet d'interpréter cette structure comme possible fosse réservée au rouissage de plantes textiles (Martial *et al.* 2004, Martial et Praud sous presse).

d'une région à l'autre (rouissoir, routoir, rutoir, roise) sont parfois encore visibles dans le paysage. C'est le cas à Lucey, près de Toul (Parc Naturel de Lorraine) où nous avons visité un terrain boisé à la sortie du village, traversé par un petit ruisseau qui alimente une trentaine de fosses alignées, encore remplies d'eau ou parfois comblées. À Lucey, chaque famille possédait sa « roise » : de forme quadrangulaire, le fond était pavé et l'un des côtés descendait en pente douce pour permettre le chargement et le déchargement du chanvre (Le chanvre au fil du temps 2006) (fig. 6).

Figure 6. Fosses aménagées en zones marécageuses près du village de Lucey (Lorraine, France). Destinées au rouissage du chanvre « familial », elles étaient encore utilisées au XXᵉ siècle (cliché : F. Médard).

Des sources ethno-historiques recensées dans le nord de la France, en Picardie (Angot 2002, Delafollie 2002, 2006) et en Lorraine (Association les Mermet 2003) témoignent de cette technique de rouissage pour le traitement du chanvre. Des fosses ou trous d'eau aménagés en zones marécageuses, alimentés par des sources ou drainés par un ruisseau, étaient situés à l'écart des villages en raison des odeurs pestilentielles dégagées par le rouissage du chanvre. Ce procédé permettait d'écourter le temps de rouissage comparé au même procédé effectué en eaux vives. En outre, « dans les routoirs, plus l'eau se renouvelait lentement, plus la filasse était solide et estimée, ce procédé était le plus utilisé » (Delafollie 2002, p. 14-15). Ces vestiges des XVIIIᵉ et XIXᵉ siècles, dont l'appellation varie

Le séchage, le battage et le teillage

Le rouissage terminé, le lin est traditionnellement lié en bottes pour le séchage. Placées debout, ces bottes sont disposées en forme de cônes pour que l'air circule entre les tiges et stoppe la décomposition (fig. 7). Il est important que les tiges soient parfaitement sèches pour que l'écorce se brise plus facilement lors de l'étape suivante : le battage. Lorsque le temps est humide, les tiges de lin peuvent être placées sur des clayettes hautes doucement chauffées par un foyer.

Le battage ou broyage sert à casser les parties ligneuses de la tige pour faire apparaître la filasse sous-jacente. Cette opération s'effectue traditionnellement à l'aide d'une broye ou en pressant les

Figure 7. Bottes de lin disposées sur pré pour le séchage
(nord de la France ?, début du XXe siècle) (carte postale ancienne).

tiges entre deux cylindres de bois. Plus simplement encore, elles peuvent être martelées sur une pierre plate à l'aide d'un battoir en bois. L'expérimentation montre que les fibres issues de tiges ainsi traitées sont souvent pliées, déchirées parfois même dédoublées lorsqu'on les observe au microscope. À l'inverse, les tiges soumises à un broyage plus doux, par exemple entre les mains, donnent des fibres quasiment intactes. Pour ne nuire ni à la solidité, ni à la longueur des fibres, il convient donc d'opter pour un broyage doux,

même si le soin apporté à l'opération se fait au détriment du rendement (Médard 2005, p. 19).

Proche de la précédente opération, le teillage permet de débarrasser complètement la filasse des débris qui y sont accrochés (parties ligneuses de la tige). Pour cela, on bat de nouveau la matière jusqu'à en éliminer les impuretés, à savoir l'épiderme des tiges (sous forme de poussières), le bois (sous forme de petits fragments appelés « anas ») et les étoupes de teillage (fibres courtes) (fig. 8).

Figure 8. Apparition de la filasse et dislocation de l'écorce au cours du battage et du teillage
(cliché : F. Médard).

73

Ce procédé a quasiment les mêmes conséquences sur la filasse que le battage : il a tendance à casser les fibres, créant parfois un déchet considérable (Kirby 1963). Le teillage peut toutefois être effectué à mains nues, méthode plus douce et moins dommageable pour la filasse. Nous savons d'expérience que cette méthode donne de bons résultats : la filasse est longue, belle, facile à filer. L'inconvénient majeur est la lenteur du procédé et l'impossibilité de traiter d'importantes quantités de matière en même temps (expérimentation réalisée en collaboration avec C. Jespersen). Or, les observations effectuées au microscope sur du matériel textile néolithique (site de Nidau BKW, BE, Suisse) révèlent une filasse de lin peu endommagée (Rast-Eicher und Thijsse 2001). La mise en perspective de ces observations tend à prouver que les Néolithiques faisaient un usage modéré des fibres de lin. En effet, si l'on souhaite traiter une matière première abondante, outillage et installations deviennent nécessaires afin de travailler plus vite en minimisant les efforts. Cela a généralement un impact négatif sur la qualité des matériaux (fig. 9).

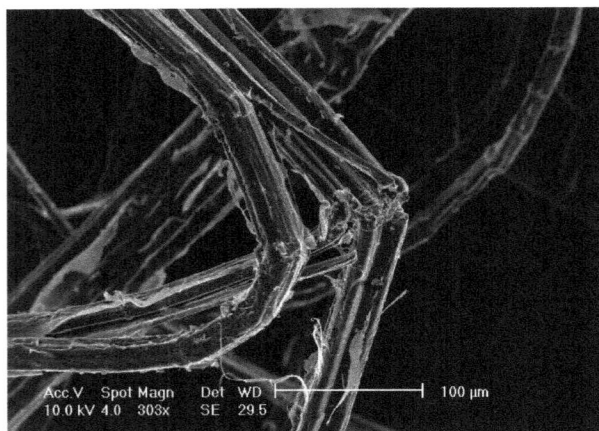

Figure 9. Fibres de lin cassées lors du teillage des tiges à l'aide d'une broye (cliché : C. Moulhérat).

L'expérimentation et l'analyse fonctionnelle livrent des résultats extrêmement intéressants sur la possibilité de teiller du lin à l'aide de simples éclats de silex. Récemment, au cours de l'étude du site d'Houplin-Ancoisne « Rue M. Dormoy », nous avons soumis à J.-P. Caspar (Caspar *et al.* 2005, sous presse) l'hypothèse d'une éventuelle utilisation d'outils en silex pour le teillage des fibres végétales, supposées préalablement rouies sur ce site (*cf. supra*), inspirée notamment par l'emploi de racloirs pour cette fonction au siècle dernier (fig. 10). Le protocole expérimental mis en œuvre consistait à racler des tiges de lin rouies

et sèches à l'aide de bords bruts de lames de silex, afin de débarrasser la filasse des fragments résiduels de l'enveloppe ligneuse.

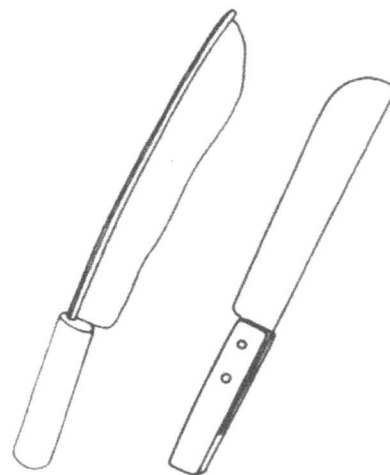

Figure 10. Racloirs utilisés pour teiller le lin (© Musée National du Lin, Courtrai, Belgique).

L'extrémité d'un faisceau de 1 à 1,5 cm de diamètre, composé d'une dizaine de tiges, est tenue dans une main pendant que l'autre racle les fibres tendues sur le bord de la lame en faisant glisser l'outil dans un mouvement souple et continu. Cette opération permet aux anas de se détacher aisément et entraîne la formation rapide d'étoupe. Ce geste doit être répété pour que les fibres soient totalement dénudées (fig. 11). Le procédé s'est avéré très efficace et peu traumatisant pour le matériau textile. L'observation des outils expéri-

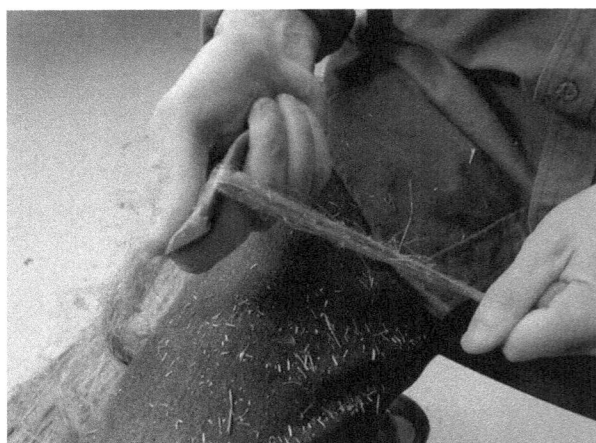

Figure 11. Teillage expérimental de fibre de lin rouies et sèches à l'aide d'un bord brut de lame en silex (expérience : J.-P. Caspar et P. Féray) (cliché : E. Martial).

mentaux sous le microscope a permis ensuite de constater la formation rapide d'un micropoli marginal, mat, souligné de stries dues à la présence de particules abrasives (terre, poussière) sur les tiges. Ce poli, par son modelé et sa texture, ressemble fortement aux traces produites par le raclage de peaux sèches à l'état souple. Les résultats de l'expérimentation amènent ainsi à réviser l'interprétation de certains polis d'usure attribués au travail de la peau et à réévaluer l'importance de l'exploitation des végétaux dans le contexte néolithique européen.

Le peignage

Le peignage consiste à démêler la filasse tout en la débarrassant des quelques déchets qui n'ont pas été réduits lors du teillage. On utilise tradition-nellement une planchette en bois hérissée de plusieurs rangées de pointes en fer, fermement maintenue sur un socle. On procède en plaçant l'extrémité d'une poignée de filasse sur les dents du peigne et en tirant vers soi. Quand la pointe est démêlée, l'artisan engage une plus grande longueur de filasse. La littérature spécifie « qu'un ouvrier trop brusque et maladroit peut occasionner un grand déchet en rompant les filaments au lieu de les démêler » (Duhamel du Monceau 1762, p. 213). Il s'agit d'une opération délicate pouvant entraîner beaucoup de perte et réduire considéra-blement la quantité de belle filasse, constituée de fibres longues.

L'observation au microscope montre que le peignage poursuit le travail effectué en amont par le rouissage : il divise les faisceaux de fibres. On obtient ainsi des faisceaux de plus en plus fins (Vétillart 1876). Un peignage bien mené permettra d'obtenir une quantité maximale de filasse longue et minimisera l'importance quantitative des étoupes et des fibres de moyenne qualité. Le soin apporté à cette étape est décisif pour les opérations ultérieures de transformation, notamment celle du filage.

Quelques artefacts néolithiques pourraient éven-tuellement être associés à cette étape du traitement des fibres, parmi lesquels une planche dans laquelle étaient autrefois fichées des pointes, aujourd'hui disparues : mise au jour sur le site de Lattrigen (BE, Suisse), elle évoque une carde ou un peigne (Vogt 1937, p. 47). Les fouilles de Feld-meilen-Vorderfeld (ZH) et d'Egolzwil (LU) ont livré chacune des faisceaux d'épines de prunelier (*Prunus spinosa* L.), dont la fonction pourrait être liée au peignage des fibres (Winiger 1981, 1995, p.165, Rast-Eicher 1990, p.119) (fig. 12).

Figure 12. Faisceau constitué d'épines de prunelier (*Prunus spinosa* L.) ligaturées, peut-être employé au peignage de la filasse. Site néolithique d'Egolzwil 3 (LU, Suisse) (Wyss 1994).

Bien qu'en l'état, l'objet ne paraisse pas maniable, l'usage du peigne à épines est mentionné dans un ouvrage traitant du filage dans les Balkans au XXᵉ siècle (Endrei 1968). L'expérimentation montre que le passage répété des épines ne réduit pas la matière première ; au contraire, il permet d'obtenir une belle qualité de filasse en générant peu de pertes.

TRANSFORMATION DE LA MATIÈRE PREMIÈRE

Les fibres sont à l'origine de toute réalisation. De leur qualité dépend celle des produits finis. L'acquisition puis la transformation des matières textiles constituent des étapes indispensables et décisives quant à la qualité et la nature des réalisations.

Le filage

La qualité des fibres joue un rôle essentiel dans les choix techniques. Pour transformer la fibre en fil, deux grandes familles techniques prédominent : le filage sans instrument et le filage au fuseau.

Le principe du filage sans instrument consiste à tordre les fibres entre les deux mains ou entre une main et une autre partie du corps (jambe, cuisse, joue …). Il compte parmi les techniques les plus répandues ; c'est la forme la plus rudimentaire de filage. Les documents ethnographiques montrent que cette méthode est adaptée au travail des fibres longues (liber) (Soustelle 1937, Seiler-Baldinger 1979), mais pas à celui des fibres courtes (lin, coton, laine …), dont la cohésion et la torsion sont difficiles à maîtriser de front. Démêlage et étirage des fibres, torsion pour former le fil, stockage du fil : ces trois étapes, propres à toutes les opérations de filage, ne peuvent être accomplies simultanément dans le cas du filage sans instrument. Chaque phase nécessite l'intervention des deux mains, à l'inverse du filage au fuseau dont la pratique favorise une plus grande liberté de mouvement permettant de travailler plus vite et d'anticiper sur la succession des gestes à accomplir.

À la différence du filage sans instrument, le filage au fuseau nécessite un axe, parfois accompagné d'une ou de plusieurs fusaïoles. Ces dernières constituent la catégorie de vestiges la mieux représentée parmi les objets issus de la culture matérielle textile (fig. 13 et 14). Ce sont des volants d'inertie destinés à alourdir et à faciliter la rotation du fuseau lors des opérations de filage. Même s'ils ne livrent pas d'informations directes sur la transformation des matières textiles, ils tendent à en préciser la qualité. En effet, l'utilisation du filage au fuseau, de par sa rapidité d'exécution, implique une fibre parfaitement propre et ordonnée n'opposant ni trop, ni trop peu de résistance. Par ailleurs, les propriétés des fusaïoles influent sur l'épaisseur et sur la résistance des fils. Elles témoignent également de l'aptitude de la matière première à subir le filage (Médard 2006a et b).

Figure 13. Fusaïoles en terre cuite du site néolithique final de Carvin (Nord, France) (cliché : E. Martial).

Figure 14. Fusaïole en pierre du site du Néolithique final de Muntelier-Dorfmatte II (FR, Suisse) (cliché : F. Médard).

Le tissage

Nos recherches ont montré que les fils néolithiques réalisés au fuseau étaient en majeure partie destinés au tissage. Le tissage s'exécute au Néolithique sur des métiers à tisser verticaux à poids. Généralement installés devant ou contre un mur, ils sont constitués de deux montants en bois, maintenus à leur extrémité supérieure par une poutre transversale dont la dimension détermine la largeur maximale de tissage. Dans le cas d'un métier incliné, l'obliquité du dispositif assure, par un jeu de gravité, la séparation des fils de chaîne en deux nappes. Deux petites fourches, fichées dans chacun des montants, servent à bloquer la barre de lisses, dont la fonction est de soulever les fils situés à l'arrière plan du métier. Enfin, une troisième barre transversale sert à maintenir à l'avant plan la moitié des fils de chaîne afin qu'ils ne se mêlent pas à l'autre moitié placée à l'arrière. Sur ce type de métier, les fils de chaîne sont montés un à un, maintenus ensemble par une lisière de départ et tendus à l'aide de pesons ; le tissage progresse du haut vers le bas (fig. 15 et 16) (Seiler-Baldinger et Médard à paraître).

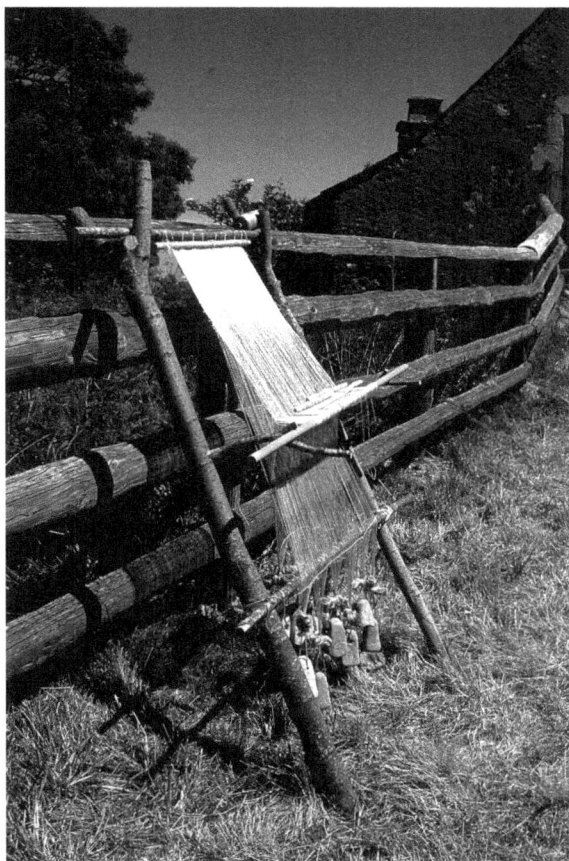

Figure 16. Pesons de métiers à tisser en terre cuite du site néolithique final de Carvin (Nord, France) (cliché : E. Martial).

En tant qu'outils employés à la confection des textiles, les pesons sont adaptés à la qualité des fils qu'ils tendent. Leur morphologie et leur poids renvoient à des besoins fonctionnels précis. Par exemple, le poids d'un peson détermine le nombre de fils qui y sont accrochés et, à ce titre, il livre des indications sur le diamètre, sur la résistance, sur les propriétés de souplesse et de rigidité des fils tissés (Médard 2006a). Ainsi, par exemple, à Raillencourt-Sainte-Olle « Le Grand Camp » (Nord, France), le comblement d'une grande fosse circulaire (4,50 m de diamètre à l'ouverture pour une profondeur de 1,30 m) comprenait un rejet de 74 pesons en terre cuite (dont 55 entiers) probablement issus du démantèlement d'un métier à tisser ; l'association de ces poids avec de gros blocs de grès n'est pas encore éclaircie (fig. 17). Au total, les six fosses de ce site daté du Néolithique final ont livré 128 pesons (dont 72 sont entiers), effectif jusque-là inédit dans le nord-ouest de la France (Bostyn et Praud 2000).

Enfin, les vestiges textiles sont instructifs, car ils permettent non seulement d'analyser les techniques de fabrication mises en œuvre mais aussi d'observer directement la matière première. Fils, cordelettes, cordes, tissus, filets, vanneries souples sont soumis à des prélèvements examinés en microscopie afin d'identifier la nature des matériaux employés (fig. 18). Grâce au développement des méthodes d'analyse, l'identification des fibres tend aujourd'hui à être systématisée. La microscopie électronique, notamment, est un outil essentiel à la reconnaissance des fibres, qu'il s'agisse de leurs caractéristiques externes (écailles, genoux de flexion...) ou internes (lumen, canal médullaire) (Médard *et al.* 2007).

Figure 15. Reconstitution d'un métier vertical à poids de type néolithique (reconstitution : M.-P. Puybaret et F. Médard) (cliché : F. Médard).

Figure 17. Rejet massif de pesons en terre cuite cylindriques à perforation verticale et blocs de grès dans une fosse du site du Néolithique final de Raillencourt-Sainte-Olle « le Grand Camp » (Nord, France) (cliché : F. Bostyn).

Figure 18. Textile néolithique du site de Feldmeilen-Vorderfeld (ZH, Suisse) (cliché : F. Médard).

CONCLUSION

Bien qu'il soit aujourd'hui encore difficile d'appréhender le processus opératoire d'extraction de la filasse de lin en Préhistoire, les recherches actuelles fournissent quelques pistes intéressantes. On s'aperçoit que la typologie ne permet pas de pousser très loin les investigations ; au contraire, dans le domaine textile, quelques associations rapides et non étayées sont probablement source d'erreur. Par exemple, nombre d'artefacts sont associés par défaut au travail des matières textiles. Des outils dont la morphologie peu spécifique ne permet pas d'identifier leur fonction initiale sont parfois attribués à la sphère de l'artisanat textile, sans véritable fondement. En revanche, les recherches effectuées dans les domaines de la tracéologie, de l'analyse des fibres textiles et de l'expérimentation fournissent des éléments intéressants.

L'expérimentation, si elle a longtemps souffert de l'amalgame effectué avec des reconstitutions fantaisistes, tant au niveau des techniques mises en œuvre qu'à celui des artefacts employés, est désormais plus structurée. Elle permet de replacer l'objet archéologique en situation fonctionnelle et de valider des hypothèses. Il est bien évident que cet axe de recherche connaît des limites : ancrée dans notre époque, l'expérimentation repose sur une pensée contemporaine et débute avec le handicap d'une perte quasi-totale des traditions techniques. Par ailleurs, les cheminements intellectuels varient d'un individu à l'autre et un résultat identique peut être obtenu à l'aide de techniques différentes. Inversement, un même geste technique effectué *a priori* suivant un seul et même protocole expérimental peut livrer des résultats différents d'un individu à l'autre ; ainsi, dans le cas d'une utilisation expérimentale d'outils en silex, l'angle d'inclinaison de l'outil sur la matière travaillée, la pression qu'exerce l'utilisateur sur l'outil, la régularité du geste effectué sont autant de critères dont la variabilité influe sur les modalités de formation d'une éventuelle trace d'utilisation. L'expérimentation, si elle ouvre des perspectives intéressantes, ne peut conduire seule à des certitudes scientifiques.

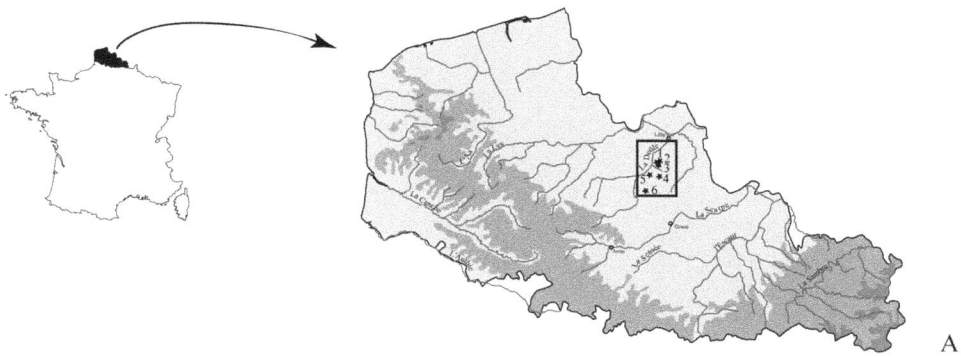

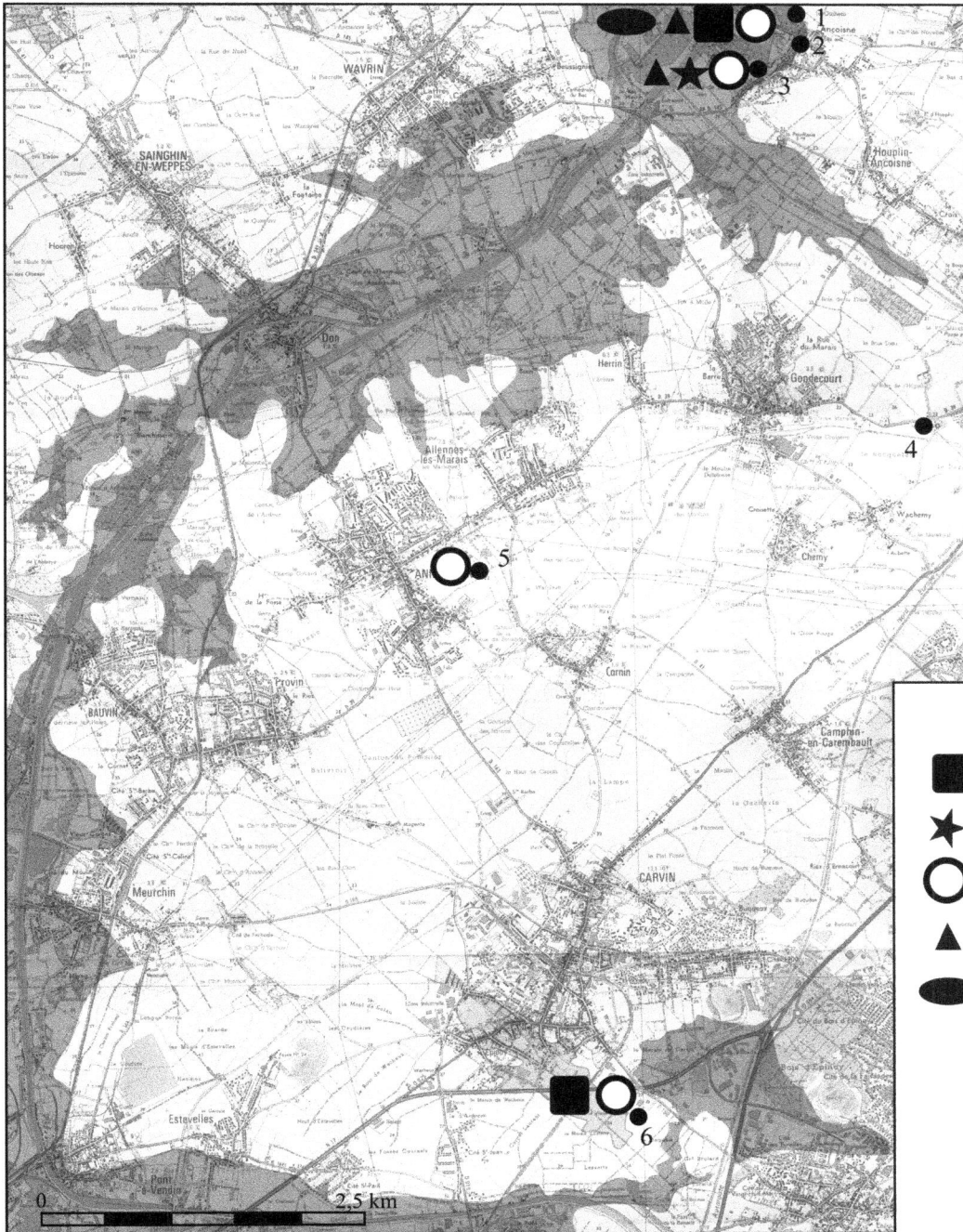

■	Peson
★	Aiguille en os
○	Fusaïole
▲	Lin
⬭	Fosse de rouissage

(d'après carte IGN 2505 O)

Figure 19. Carte de répartition des traces d'artisanat textile au Néolithique final dans la vallée de la Deûle (Nord, France). 1 : Houplin-Ancoisne « Rue M. Dormoy » ; 2 : Houplin-Ancoisne « Rue G. Peri » ; 3 : Houplin-Ancoisne « Le Marais de Santes » ; 4 : Seclin « Les Euwis » ; 5 : Annoeullin « Rue Lavoisier » ; 6 : Carvin « Z.I. du Château ».

La documentation ethnographique permet d'observer les outils et les gestes qui les accompagnent dans leur contexte fonctionnel. Elle permet d'appréhender des savoir-faire différents, ouvre des perspectives nouvelles et élargit le champ des investigations archéologiques. Cependant la comparaison des documents ethnographiques et archéologiques doit être maniée avec prudence : la similitude entre des objets passés et contemporains ne signifie pas qu'ils ont été conçus selon les même techniques et pour la même finalité. Il convient donc de garder une certaine distance avec l'utilisation de ces données (Médard et al. 2007).

La pluridisciplinarité reste la meilleure manière de répondre aux interrogations dans le domaine des matières textiles préhistoriques. En effet, on ne peut pas compter sur les seuls vestiges matériels pour appréhender l'existence d'activités textiles sur un site pré- ou protohistorique : nous avons vu que la plupart des étapes d'acquisition et de transformation des matières textiles d'origine végétale se déroulent sans outils. Quelques-uns sont fabriqués dans des matériaux organiques qui, tout comme les restes textiles proprement dits, ne sont conservés qu'exceptionnellement en contexte humide et/ou anaérobie. L'absence de preuves matérielles en contexte archéologique n'est pas preuve de l'absence... Une première approche spatiale des traces d'artisanat textile sur les sites du Néolithique final du nord de la France (Martial 2006) semble indiquer une chaîne opératoire géographiquement fractionnée, probablement au sein d'un territoire dont elle pourrait refléter l'organisation (fig. 19).

Ainsi, l'omniprésence des fusaïoles semble concorder avec le lourd investissement en temps et en main d'œuvre indispensable à la production de kilomètres de fil pour la confection d'étoffes tissées. En revanche, la distribution des pesons de métiers à tisser et celle des restes archéobotaniques de lin semblent montrer que la culture de cette plante, d'une part, et l'exécution du tissage à l'autre extrémité de la chaîne opératoire, d'autre part, sont réservées à certains sites. Tout en ayant à l'esprit la relative pauvreté de la documentation archéologique (les sites ne sont jamais fouillés in extenso et la perte d'information due aux conditions de conservation est énorme en milieu terrestre), il nous semble intéressant de mener une réflexion sur l'organisation territoriale de l'artisanat textile au cours du Néolithique en se fondant sur une approche pluridisciplinaire.

ADRESSE DES AUTEURS

Emmanuelle MARTIAL
UMR 7041 (Protohistoire européenne)
Inrap Nord-Picardie
11, rue Victor Hugo
59350 Saint-André – France
emmanuelle.martial@inrap.fr

Fabienne MÉDARD
UMR 7041 (Protohistoire européenne)
Maison de l'archéologie et de l'ethnologie
21, allée de l'Université
92023 Nanterre Cedex – France
fabienne.medard@wanadoo.fr

REMERCIEMENTS

Nous remercions Marc Talon pour les articles sur le rouissage du chanvre en Picardie qu'il nous a aimablement fournis ainsi que Kai Fechner qui nous a accompagnés lors de notre visite des « roises » à Lucey. Nous adressons également nos remerciements à Michael Ilett qui a assuré, avec diligence, la traduction anglaise des documents qui nous ont été nécessaires.

HOMMAGE

Cet article est dédié à Jean-Paul Caspar, qui vient de nous quitter. Le sujet traité ici alimente une réflexion que nous avions activement partagée et qui l'avait incité à réévaluer l'importance accordée au traitement et à l'exploitation des végétaux au cours du Néolithique en Europe. Il était lui aussi convaincu de l'intérêt d'une approche pluridisciplinaire avec toute la richesse d'échanges, d'idées et d'expériences qu'apporte le mélange des genres et des cultures à la connaissance des sociétés humaines.

Emmanuelle Martial, le 11 septembre 2007.

BIBLIOGRAPHIE

ANGOT, J.-P., 2002. Le lin et le chanvre dans le Noyonnais. *Revue semestrielle de la Société Historique, Archéologique et Scientifique de Noyon, Ét*udes noyonnaises, 267, 26-29.

ASSOCIATION LES MERMETS, 2003. Les Mermet d'hier et d'aujourd'hui. Profession : peigneur de chanvre. Available from: http://www.mermet.asso.fr (Accessed 2003).

BONNÉTAT, L., 1919. Les plantes textiles. Lin, chanvre, etc. Paris : Hachette.

BONNIER, G., nd. Plantes médicinales, plantes mellifères, plantes utiles et nuisibles. Paris : Librairie générale de l'Enseignement (édition ancienne).

BOSTYN, F. et PRAUD, I., 2000. Le site néolithique de Raillencourt-Sainte-Olle « Le Grand Camp » (Nord). Internéo, 3, 119-129.

CASPAR, J.-P., FERAY, P. et MARTIAL, E., 2005. Identification et reconstitution des traces de teillage des fibres végétales au Néolithique. Bulletin de la Société préhistorique française, 102 (4), 867-880.

CASPAR, J.-P., MARTIAL, E. et FERAY, P., sous presse. Le teillage des fibres végétales : pour une réinterprétation fonctionnelle d'outils en silex néolithiques. In : Actes du 26e Colloque interrégional sur le Néolithique, 8-9 novembre 2003, Musée National d'Histoire et d'Art, Luxem-bourg. Luxembourg : Archeologia Mosellana.

DELAFOLLIE, M., 2002. Le rouissage du chanvre au XIXe siècle. La revue du pays d'Estrées, 3, 14-16.

DELAFOLLIE, M., 2006. Les routoirs de Bazicourt. La revue du pays d'Estrées, 20, 3-10.

DIETSCH-SELLAMI, M.-F., DURAND, J. et VERDIN, P., 2006. Traitement des plantes textiles et prépara-tion alimentaire à Meaurecourt « La Croix de Choisy » (Yvelines) au Néolithique ancien : questionnement à partir des premiers résultats. In : Archéologie des textiles et teintures d'ori-gine végétale. Archéologie des fourrages, Archéobotanique 2006, Compiègne 28-30 juin 2006. Pré-actes.

DUHAMEL DU MONCEAU, H.-L., 1762. Éléments d'agriculture, vol. 2. Paris.

DURAND, S., BEMILLI, C., BONNARDIN, S., DIETSCH-SELLAMI, M. F., DURAND, J., HAMOND, C., MAIGROT, Y., PRAUD, I. et VERDIN, P., 2006. Présentation du site néolithique ancien de Maurecourt « La Croix de Choisy » (78). Internéo, 6, 19-32.

ENDREI, W., 1968. L'évolution des techniques du filage et du tissage du Moyen-Âge à la révolution industrielle. Paris : Mouton.

EWERS, M., 1989. Linum usitatissimum L. Le lin, une plante cultivée du Néolithique. Bulletin de la Société Préhistorique Luxembourgeoise, 11, 169-192.

KIRBY, R. H., 1963. Vegetable fibers. London: Leonard Hill (books) Ltd.

LE CHANVRE AU FIL DU TEMPS, 2006. Le chanvre au fil du temps. Maison Lorraine de Polyculture : Édition Parc naturel régional de Lorraine.

LEFRANC, P. et ARBOGAST, R. M., 2000. L'habitat néolithique moyen et récent de Holtzheim « Zone d'activités - phase 3 » (Bas-Rhin). Internéo, 3, 59-69.

LES LACUSTRES, 150 OBJETS RACONTENT 150 HISTOIRES, 2004. Les Lacustres, 150 objets racontent 150 histoires. Catalogue d'exposition. Zurich : Musée National Suisse.

LIEUTAGHI, P., 1998. La plante compagne. Pratique et imaginaire de la flore sauvage en Europe occidentale. Arles : Actes Sud.

MARTIAL, E., avec la collaboration de Boulen, M., Caspar, J.-P., Coubray, S., Sellami, M.-F., Fechner, K., Feray, P., Maigrot, Y. et Praud, I., 2006. Les témoins de l'exploitation des végétaux pour l'artisanat textile au Néolithique final, dans le Nord de la France. In : Archéologie des textiles et teintures d'origine végétale. Archéologie des fourrages, Archéobotanique 2006, Compiègne 28-30 juin. Pré-actes.

MARTIAL, E., PRAUD, I. et BOSTYN, F., 2004. Recherches récentes sur le Néolithique final dans le Nord de la France. In : M. Van der Linden et L. Salanova, eds. Le troisième millénaire dans le Nord de la France et en Belgique. Mémoire de la Société préhistorique française XXXV, Anthropologica et Praehistorica, 115, 49-71.

MARTIAL, E. et PRAUD, I. avec la collaboration de Boulen, M., Braguier, S., Caspar, J.-P., Clavel, B., Coubray, S., Deschodt, L., Dietsch-Sellami, M.-F., Fechner, K., Lehnebach, C. et Maigrot, Y., sous presse. Un site palissadé du Néolithique final à Houplin-Ancoisne (Nord, France). In : Actes du 26e Colloque interrégional sur le Néolithique, 8-9 novembre 2003, Musée National d'Histoire et d'Art, Luxembourg. Luxembourg : Archeologia Mosellana.

MÉDARD, F., 2004. Identification des matières textiles préhistoriques et des traitements mis en œuvre pour les obtenir. Exemples du lin et de l'ortie. Rapport intermédiaire d'activité scientifique, inédit. Fond National Suisse.

MÉDARD, F., 2005. Préparation et transformation du lin destiné à la production des fils extrêmement fins. Données archéologiques, anatomiques et expérimentales. Bulletin de liaison du CIETA, 82, 6-24.

MÉDARD, F., 2006a. Les activités de filage au Néolithique sur le Plateau suisse. Analyse technique, économique et sociale. Monographie du CRA, 28. Paris : Éditions du CNRS.

MÉDARD, F., 2006b. La fusaïole : au delà des idées reçues... In : L. Astruc, F. Bon, V. Léa, P.-Y. Milcent et S. Philibert, eds. Normes techniques et pratiques sociales. De la simplicité des outillages pré- et protohistoriques. Antibes : Éditions APDCA, 275-280.

MÉLARD, F., sous presse. L'activité textile sur le site de Pfäffikon-Burg (ZH). *In:* U. Eberli, ed. *Seeu-fersiedlungen. Pfäffikon-Burg.* Zürich und Egg : Zürcher Archäologie.

MÉLARD, F., MICOUIN-CHEVAL, C. et MOULHÉRAT, C., 2007. Vestiges et artefacts textiles pré et protohistoriques : historique des recherches et nouvelles approches. *In : Actes du colloque du centenaire de la SPF, sept. 2004, Avignon.*

PLINE L'ANCIEN, 1964-réédition. *Histoire Naturelle, XIX.* Clermont-Ferrand : Les Belles Lettres.

RAST-EICHER, A., 1990. Die Verarabeitung von Bast. *In: Die Ersten Bauern, Band 1.* Zürich: Schweiz, 119-122.

RAST-EICHER, A. und THIJSSE, S., 2001. Anbau und Verarbeitung von Lein: Experiment und archäologisches Material. *Zeitschrift für Schweizerische Archäologie und Kunstgeschichte*, 58 (1), 47-56.

REINHARD, J., 2000. Les poids de tisserands. Textiles et vannerie. *In :* D. Ramseyer, ed. *Mantelier-Fischergassli, un habitat néolithique au bord du lac de Morat.* Archéologie fribourgeoise, 15. Fribourg : Éditions Universitaires, 193-205.

SAUDINOS, L., 1942. *L'industrie familiale du lin et du chanvre.* Toulouse : Librairie Edouard Privat.

SEILER-BALDINGER, A., 1979. « Händgematten-Kunst ». *In: textile Ausdrucksform bei Yaguaund Ticuna-Indianern Nordwest-Amazoniens.* Basel: Verhandlungen der Naturforschenden Gesellschaft, 90, 61-130.

SEILER-BALDINGER, A. et MÉDARD, F., à paraître. Les textiles cordés : armures et techniques. *Techniques et Cultures.*

SOUSTELLE, J., 1937. La famille Otomi-Pame du Mexique Central. Paris : Institut d'Ethnologie.

VETILLART, M., 1876. *Études sur les fibres végétales textiles employées dans l'industrie.* Paris : Librairie de Firmin-Didot et Cie.

VOGT, E., 1937. *Geflechte und Gewebe der Steinzeit.* Basel: Monographien zur Ur- und Frühgeschichte der Schweiz 1.

WINIGER, J., 1981. *Feldmeilen-Vorderfeld. Der Übergang von der Pfyner zur Horgener Kultur.* Antiqua 8. Basel.

WINIGER, J., 1995. Die Bekleidung des Eismannes und die Anfänge der Weberei nördlich der Alpen. *In: Der Mann im Eis. Neue Funde und Ergebnisse, Band 2.* Wien: Springer Verlag, 119-187.

WYSS, R., 1994. *Steinzeitliche Bauern auf der Suche nach neuen Lebensformen. Egolzwil 3 und die Egolzwiler Kultur, Band 1.* Zürich: Archaeologische Forschungen.

INVESTIGATING SOCIAL ASPECTS OF TECHNICAL PROCESSES: CLOTH PRODUCTION FROM PLANT FIBRES IN A NEOLITHIC LAKE DWELLING ON LAKE CONSTANCE, GERMANY

Susanna HARRIS

Abstract: Plants were an important source of raw materials for many types of cloth in Europe during the Neolithic. On the basis of evidence from waterlogged sites, the most significant raw materials were flax and tree bast, which were used to create woven textiles, twined cloth and netting. Looking towards the investigation of cloth in social anthropology, I propose that issues of time and place are a significant aspect of the social context of cloth. The aim of this paper is to investigate these issues in relation to processing cloth from plants, through a case study of the Neolithic lake dwelling of Hornstaad Hörnle IA on Lake Constance, Germany, dated ca 3900 BC. Building on recent research into the way these cloth types were processed and constructed in the Neolithic, I use the evidence at Hornstaad Hörnle IA to consider the location of these tasks in the home, village and wider landscape, as temporal aspects carried out daily, seasonally or over a number of years and in relation to the lives of members of the village. Through this I hope to show how the methods of processing cloth were not only a technical solution to satisfy the need for various types of cloth, but were part of the way that people's lives were created and made meaningful.

Keywords: cloth, process, social life, Neolithic, time, place.

Résumé : Les plantes sont une importante source de matière première, permettant la fabrication de nombreux types de tissus en Europe, durant le Néolithique. À partir des découvertes réalisées dans les sites de milieu humide, il a pu être établi que les matériaux les plus significatifs sont le lin et la filasse d'écorce qui ont été utilisés pour la confection de textiles tissés, de tissus tressés et de filets. Dans la perspective des recherches menées en anthropologie sociale, je propose d'étudier le contexte social de production des textiles à travers les concepts de temps et d'espace. L'étude de cas concerne le site néolithique lacustre de Hornstaad Hörnle IA du lac de Constance (Allemagne), daté de 3900 av. J.-C. À partir des données bibliographiques disponibles concernant les techniques de fabrication des différents tissus au Néolithique, j'analyse la localisation des différentes opérations menées que celles-ci se déroulent à l'intérieur des maisons, dans le village ou au-delà et j'étudie la périodicité de ces opérations qu'il s'agisse de tâches quotidiennes, saisonnières ou se déroulant sur plusieurs années, en relation avec le cycle de vie des membres du village. Par ce biais, j'espère montrer en quoi la production de tissus ne constitue pas uniquement une solution technique visant à satisfaire des besoins variés mais fait aussi partie de la façon dont les gens vivent et donnent sens à leur vie.

Mots-clés : tissus, chaîne opératoire, vie sociale, Néolithique, temps, espace.

INTRODUCTION

Current research on plant processing in archaeology is increasingly providing accurate evidence for the technical practices, typology and quantitative significance of plants in prehistoric societies. This detailed understanding of material and artefacts is essential, however additional questions need to be asked to comprehend the significance of this evidence for understanding past societies. This article examines Neolithic plant processing to make cloth from a social perspective, based in the consideration of time and place as aspects of social life. The overall aim of this approach is to take the evidence of plant processing to produce cloth into a realm that is tangibly human; focusing on the people who were engaged in making the artefacts, rather than the processes or resulting artefacts alone. This approach is rooted in interpretative material culture and technology studies that argue for the importance of investigating technologies, like material culture, as social phenomena (Pfaffenberger 1988, Ingold 1993, Lemonnier 1993, Dobres 2000).

To approach the evidence in this way, I shall focus on time and place as aspects of process and social life. By time, I mean the passage of time during a human lifetime, over a number of days or through the seasons; by place I mean the location of tasks whether inside a house, around a settlement, in the forest or fields. In the first part of this article, I present the case for the social relevance of these aspects of time and place, and propose these concepts as a method of investigating social aspects of technical processes. In the second part, I examine the archaeological evidence for the technical processes used to make cloth from plants in the Neolithic with a case study of the lake dwelling of Hornstaad Hörnle IA, Lake Constance, Germany. I then consider the aspects of time and place that relate to these processes and combine the evidence from the case study with ethnographic and historical sources. In the results, I reflect on the implications that the time and place of processing cloth from plants may have had on the live of the inhabitants of Hornstaad Hörnle IA.

This research is based on bibliographic sources, the excavation reports and post-excavation analysis of Hornstaad Hörnle IA combined with published research of plant processing in the Neolithic, as well as further historical, ethnographic and experimental information on similar practices of plant processing. Through this combination of sources I have brought together a wide range of evidence for the time and place of plant processing to make cloth in the Neolithic, focusing on the case study. Some of these sources provide specific evidence for time and place, for example the location of fields in relation to the lake dwelling of Hornstaad Hörnle IA. Others are more general, such as the amount of time it could take to spin thread by twisting on the thigh based on an ethnographic example.

Time, place and social life

Time and place include questions such as " how long? ", " when? " and " where? ". These questions are useful to archaeologists, because these aspects are common to all processes. However, it is necessary to consider why time and place are inherently social aspects of technical process.

Processes and " *taskscape* "

From a theoretical perspective Ingold (1993) describes the social nature of the time and place in carrying out tasks. He uses the term " *taskscape* " to refer to the complete range of related activities performed by a society. Practised over a number of days, years, a lifetime, whether in the home or in fields surrounding the home, alone or with others, these activities have a context of time and place and are part of a individual's normal working and social life. The activities of such a *taskscape* are not performed in a vacuum, but are the way people interact and are part of society. " The temporality of the *taskscape* is social, then,

not because society provides an external frame against which particular tasks find independent measure, but because people, in performance of their tasks, also attend to one another " (Ingold 1993, p. 518). The combination of activities in the *taskscape,* the way they are carried out, perceived and understood are characteristic of the people and societies that perform them.

A time and place for learning and developing skills

These *taskscapes* may vary for different members of the community, and develop and change throughout an individual's lifetime. From this perspective, the coincidence of task, time and place occurs in different ways throughout an individual's lifetime. For example, in the community of Chinchero, Peru, spinning and weaving is a woman's task that is learned and practised throughout her life (Franquemont and Franquemont 1987). From a young age, girls learning to spin at home with their mother, later while looking after sheep, they will learn to weave from their peers. More complex weaving is learnt at home from a more experienced woman when the girl is older. When she gets married, these skills are put into use to set up her home. As wife and mother, there is usually little time to spin and weave, but once the children leave home she has time to spend at home at the loom, and in old age may significantly develop these skills, until her eyesight or dexterity fails and she again returns to spinning. These meetings and relationships, associated with age and place, are an essential part of learning and developing skills to work materials, as well as conforming to expectations of gender and age roles. Conversely, conflicting cycles of time and place may prevent these opportunities arising.

Symbolism and beliefs

Another distinctive aspect of technical processes in societies is the way they are imbued with symbolic significance and beliefs. This is explored in M. J. Adams' article (1971) based on ethnographic fieldwork in a village of Indonesia. In this paper she describes the time schedules for producing ceremonial cloth (woven from cotton grown by the villagers), including the annual seasonal cycle of agricultural work, the length of time to produce a cloth (up to four years), and the relationship of these to life-cycle events. These processes are intricately related to symbols of male and female, mythology, and processes of social life, including the meeting of young people, pregnancy and the negotiation of in-law relationships. In this example, the author shows many ways that the time cycle of processing cloth is bound to symbolic beliefs as much as the finished cloth is. This is just one example of the many ways that aspects of processing materials can be imbued with symbolic meaning and associated with beliefs in societies.

DEVELOPING A METHOD

These examples described above show some of the ways that time and place are socially relevant in the tasks of processing materials. To investigate the archaeological evidence in this way, the first step is to identify the processes that were carried out in the past. In the case of plant processing for cloth in the Neolithic, there is a substantial body of current research relating to archaeological evidence with processes known from historical, ethnographic and experimental sources to understand how the processes were carried out in the past (Körber-Grohne und Feldtkeller 1998, Médard 2000, Moser et Médard 2001, Bazzanella *et al.* 2003a, Médard 2003, Feldtkeller 2004). From this it is then necessary to consider the aspects of time and place that relate to these processes, evaluating these results in relation to a specific case study. The aim of this method is to take the technical process of plant processing to make cloth in the Neolithic and investigate it from a social perspective. To do this, I now turn to the example of Hornstaad Hörnle IA, a lake dwelling

on the shore of Lake Constance, Germany, dated to the early fourth millennium.

CASE STUDY: THE LAKE DWELLING OF HORNSTAAD HÖRNLE IA

Hornstaad Hörnle IA is a waterlogged lake dwelling settlement, situated on the shore of the Hori Peninsula of the Untersee on Lake Constance (fig. 1 and 2). The site was excavated by H. Schlichtherle (1990) between 1973-1980, and B. Dieckmann (1987, 1991) between 1983-1993 (see also Maier 2001). It is dated by dendrochronology between 3917 and 3905 BC (Billamboz 1998) and is the older of two Recent Neolithic (*Jungneolithikum)* lake dwellings built on the same site. I shall only deal with the first one (Hornstaad Hörnle IA) in this paper.

The stratigraphy of Hornstaad Hörnle IA shows three main phases. The first (3917-3910/3909 cal. BC) and last (3909-3905 cal. BC) are represented by loam, sand and uncarbonised organic layers of waste materials. The second layer is a burnt layer, representing the destruction of the village by fire in late summer or early autumn of 3910/3909 cal. BC. The burnt layer includes large quantities of charred wood and cereals and most of the preserved cloths (as catalogued in Schlichtherle 1990). Wooden building posts are present in all layers. The houses of the lake dwelling were made of timber posts and built roughly in rows, they are approximately 3,5 m wide and 9 m long (fig. 3).

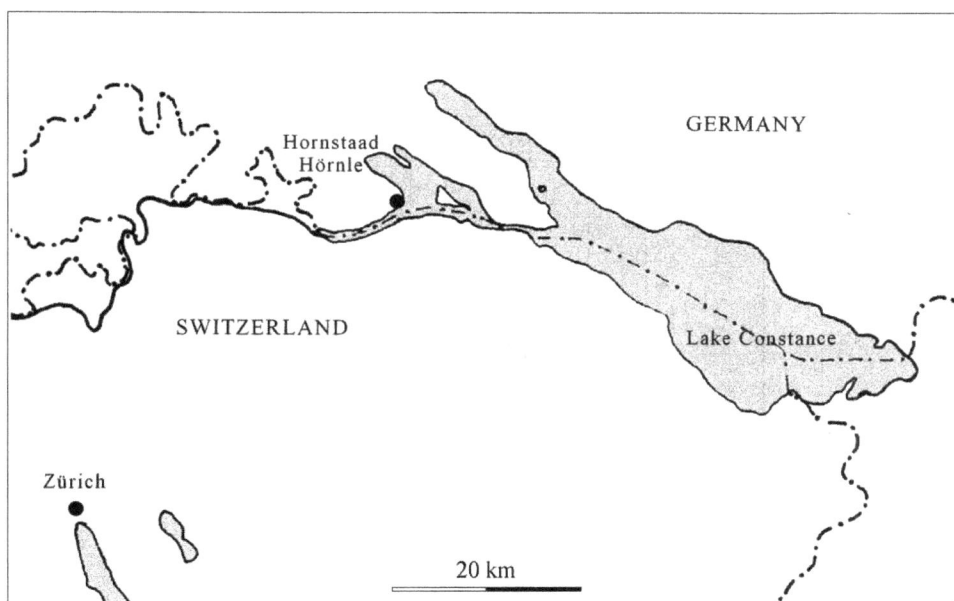

Figure 1. Location map of Hornstaad Hörnle, Lake Constance, Germany.

There were around 40 houses, probably representing the homes of several hundred people (Dieckmann 1991, Maier and Vogt 2000).

Village life

As would be expected in the Neolithic, the inhabitants cultivated crops and kept domestic animals, however hunting, fishing and collecting forest resources were important. Only about 30 % of the animal bones from Hornstaad Hörnle come from domestic animals, mainly cattle, with over 50 % from wild animals mostly red deer, and around 15 % fish (Kokabi 1990). Cereals, flax, poppies and peas were cultivated by the villagers and stored in their houses (Maier 1999, Maier and Vogt 2000). The villagers collected wild fruit, berries and nuts (Maier 1990), and tree bast for fibres. Some materials were imported, for example, flint and stone (Hoffstadt und Maier 1999) and there is noticeable production of perforated limestone beads in some areas of the village (Dieckmann 1991).

Evidence for cloth and plant processing for cloth

There are over 1500 cloth fragments excavated from Hornstaad Hornle IA (Müller 1994) and two

analyses of the cloth finds, performed by Schlichtherle (1990) and Körber-Grohne and Feldtkeller (1998). The first one is an analysis and catalogue of finds from the earlier excavations within a monograph of the site; the latter is the work of botanical and textile specialists with analysis of the raw materials, fibre processing treatments and construction types. These reports provide the evidence for plant species, some of the aspects of processing fibres and the type of cloth construction at Hornstaad Hörnle IA. As discussed above, additional bibliographic sources are used on plant processing, spinning and cloth construction in the Neolithic to understand more about the plant processing techniques that were used at the time of the settlement. When necessary, these are compared with historical, experimental or ethnographic sources.

The main cloth construction types from the lake dwelling settlement of Hornstaad Hörnle IA are numerous variations of twined cloth and netting, plus a few woven textiles. The raw materials are mainly tree bast (majority lime, some oak and elm) and flax, as identified through microscopic examination of the cell structure. The cloth here is like the cloth types at other Neolithic lake dwelling sites (Rast-Eicher 2005), although each site is unique in some aspects.

Figure 2. The site of Hornstaad Hörnle in 2005 (Photograph by S. Harris).

In addition to the cloth artefacts at Hornstaad Hornle IA, there is extensive analysis of the botanical remains, including evidence of fibres processing (Maier 2001) and a soil analysis to investigate the use of the surrounding landscape for cultivation and forest resources (Vogt 1990, Maier 1999, Maier and Vogt 2000). Although there is not yet a full analysis of the distribution of cloth finds according to the house plans, there is a spatial analysis of the nets and net-weights in relation to houses (Dieckmann 1991).

Most of the cloth finds are fragments, so it is not possible to establish how they were used, however, in some case the use can be identified. The evidence of large fragments of nets, weights and sinkers associates the nets with fishing. There are examples of basketry containers with mesh bases made from twining technique that were probably sieves. A tufted cone-shaped object is interpreted as a hat; and a small plain-weave linen draw-string object is recognised as a bag (Schlichtherle 1990, Bazzanella *et al.* 2003b).

FROM PLANTS TO CLOTH

I shall consider what we can understand of the processes carried out at Hornstaad Hörnle IA to make these different types of cloth from flax and tree bast. I follow the process from the acquisition of raw materials to finished cloth, paying particular attention to issues of time and place.

Flax cultivation

The analysis of flax plant and seeds remains shows the plants are characteristic of domestic flax (*Linum usitatissimum*). Flax is an annual plant and would need to be cultivated each year. The villagers cultivated flax throughout the occupation of Hornstaad Hornle IA; there is evidence for flax before and after the fire and in the burnt layer (Schlichtherle 1990, Maier 2001). Based on the pedological survey and the reconstruction of Neolithic soils, it is proposed that the village cultivated a large field area on the drier, fertile soils away from the lake shore, relatively close to

Figure 3. Possible reconstruction of the house type at Hornstaad Hörnle IA
(Drawing by S. Notaro after a model *in* Konstanz Ländesmuseum).

87

the village, within a radius of 300-700 m from the settlement (Vogt 1990, Maier 1999, Maier and Vogt 2000, Maier 2001). To cultivate the land, the settlers would have cleared the original mixed forest. Comparison from the examination of woodland weeds in the cereals and flax suggests that flax may have been cultivated closer to pioneer woodland than the cereals were.

Field cultivation methods are part of a wider debate in archaeology (Rösch et al. 2002, Bogaard 2004). Taking a site-specific approach, the lack of woodland plants, perennial weeds and dominance of annual seeds in the cereal store of Hornstaad Hörnle IA suggest that after the initial clearance the fields were cultivated continuously and worked intensively by digging and hoeing over many years without fallow periods (Maier 1999). However, the same evidence is used to argue that shifting slash and burn cultivation methods could have created similar results, as fire acts to suppress weeds (Rösch et al. 2002).

For Maier (1990, 1999, 2001), it is debatable if cereal crops were sown in summer or winter as both are possible, the same may be relevant to flax. In terms of the harvest time, concentrations of partially broken flax seeds, interpreted as remains of faeces rather than lost seeds indicate that flax seeds were eaten while the fibres were processed from the stem. In this case, the flax plants would need to be harvested late in the summer when the seeds were fully developed.

The remains of preserved flax stems with roots indicate that, at harvesting, flax was pulled up by its roots rather than cut. This would be best practice for the exploitation of fibre, as it retains the maximum length of the fibres. In addition, it is difficult to harvest flax by cutting; whereas it is comparatively easy to pull up the plant by hand as the roots are shallow (V. Beugnier pers. comm.). In terms of growing and harvest season, flax falls into a similar time/place structure to cereal and poppy cultivation (Maier 1990).

Collecting tree bast

Tree bast is the inner bark, found beneath the outer bark and removed when the bark is stripped from the wood. Lime, oak and elm were part of the wider tree flora and are represented in the pollen diagrams (Rösch 1996). Based on the analysis of soil types and the archaeobotanical remains, Maier and Vogt (2000) offer a model of possible areas where these bast-providing tree species grew: lime,

oak and elm would have occurred beyond the cultivated area on the luvisols; the wet areas near the shore would have only supported willow and poplar, while the areas beyond the flood zone probably supported some oak and elm.

Based on historical examples, the easiest time to remove bast is in early spring when the sap is flowing, which was then followed by water retting (see below) until autumn (Körber-Grohne und Feldtkeller 1998, Myking et al. 2005). However as the tree bast at Hornstaad Hornle IA does not seem to have been water retted, it may have been collected at a different time. Alternative processes known historically from Scandinavia account for a late autumn or winter collection of branches for bast (op. cit.). Therefore, at Hornstaad the seasonal collection of tree bast remains unclear, but could have been spring or late autumn/winter. If collected in autumn, the task of gathering tree bast may have coincided with the collection of seasonal fruit and berries including apples, blackberries, hazelnuts, beech nuts and sloes, from the surrounding woodland (Maier 1990) (fig. 5).

In terms of the age of trees or branches exploited for bast, the finest extracted fibres used for thread are only 1-2 mm wide and two to three year-rings thick; the narrowness and fineness of the fibres suggests that young trees and branches were used. There are also thicker fibres with numerous year-rings (Körber-Grohne und Feldtkeller 1998). As bark does not grow back, continuous exploitation of bast depends on the growth of new trees and branches.

Whether the woodland was managed by methods such as coppicing remains unclear. Based on the dendrotypological evidence of the wood posts used in the lake village it is possible that there were periods of regeneration in woodlands following periods of felling that may be equated with coppicing (Billamboz 1990).

Despite the large quantities of wood used for building the lake dwelling there is very little lime or elm for building, in contrast to the high quantity of oak (Schlichtherle 1990). This contrasts to the proportions of bast fibres, where lime is the most common. It would appear that these processes were not closely interconnected, which may have been for several reasons, for example: lime is a poor wood for building, bast was possibly removed from trees that were too young to be useful for building, or possibly bast was removed from a living tree as opposed to a felled tree.

Processing flax and tree bast to extract fibres

Flax

Once the flax plants are harvested, the stems are ready to be stored or processed immediately to extract the fibres. Two bundles of whole charred flax plants in the area of Houses 10 and 11 suggest the plants were being stored in the houses when the village burnt down (Maier 2001), as would be expected if the occupants were storing dried flax plants to process for fibres. The first stage of the processing is to remove the seeds. In farming communities in the 19th century, the seed capsules were combed from the stems (*rippling*) after the plants were dried, then later in the autumn crushed to remove the seed and sieved. Several concentrations of broken seed capsules in the occupation layers before and after the fire are thought to represent this process.

From the microscopic examination of the flax fibres and fibre bundles that were spun into thread, the missing stem-epidermis suggests that the stems had partially rotted in water (*retting*) (Körber-Grohne und Feldtkeller 1998, Maier 2001). In comparison to historical processes, it is likely that the bundles of flax stems were submerged in water (that same autumn?) for several weeks to remove the outer-wall of the stem, leaving the central woody area and fibres (Flad 1984). The lake could have been a suitable place for retting.

After retting and subsequent drying the stems are then ready for breaking (*bracking*), that involves snapping the central woody part, and the next step is scraping (*scutching*) to remove the unwanted stem parts. Although no tools have been identified in relation to this process at Hornstaad Hörnle IA, the remains of the broken woody part of flax stems, " scutching debris " (ca 1-5 cm long), found in 27 areas of the excavation, mainly the occupation layer after the fire, are evidence that the flax plants were being broken to allow the fibres to be extracted in the village (Maier 2001). After this the fibres were separated by combing (*heckling*), which is evident through microscopic examination of the fibres (Körber-Grohne und Feldtkeller 1998). This presumably occurred in the village although the process leaves no archaeological evidence beside the fibres themselves.

Tree bast

There are unprepared tree bast strips in all layers of the excavation; however, these may either represent waste product for fibre production or tree bast stored in a crude state to be worked later (Schlichtherle 1990). There is a knotted bundle of tree bast strips (Müller 1994), which may be a stored hank of bast fibres. In some cases the tree bast was probably used unprepared with the bark removed, for example to tie on net-weights. The question remains of how the fibres were rendered supple enough for the production of cloth.

Tree bast can be processed by water retting, like flax. However, as observed from microscopic examination of the cloth artefacts by Körber-Grohne and Feldtkeller (*op. cit.*), the annual layers of tree bast at Hornstaad Hörnle IA remain attached, as do the storage cells, suggesting the bast was prepared differently. As mentioned above, historical treatment methods from northern Europe, researched by Feldtkeller suggest that the bark could be removed from felled branches in late autumn by heating over a fire or in an oven. The dampness in the wood would force the bast to split from the bark so that it could be peeled and scraped off; the bast would then be rubbed and agitated between the fingers like a rag to separate the fibres. Fledtkeller and Körber-Grohne suggest that this or a similar method is possible at Hornstaad Hörnle IA, although the exact process remains unclear.

From fibre to thread: plaiting, twisting and spinning

The villagers were using a range of thread types from plant fibres, including unspun or weakly spun strips of tree bast, simple-spun (single thread) and plied threads (two threads spun together), in fine and coarse versions. More rarely, tree bast threads in the passive system (warp) are narrow plaits made of single twisted threads (Schlichtherle 1990). F. Médard (2003) has investigated the spinning methods used in the Neolithic and suggests two main methods: spinning without a spindle by twisting threads on a surface, such as the thigh, and spinning with the use of a spindle.

Hand spinning by twisting on the thigh

As there is no equipment for this method of spinning, the spinning technique is assumed on the basis of the finished thread characteristics. According to F. Médard (*op. cit.*) this technique is most suited to long fibres, such as tree bast, and for producing threads over 1 mm in diameter. Through examining the fibres and comparison with ethno-graphic methods, she believes that this method was used in the Neolithic. If this was the case, it was probably practised at Hornstaad Hörnle IA to spin tree bast thread for twined cloth and cords.

Spinning with a spindle

The use of a twisting device, such as a spindle with a spindle whorl (central weight threaded onto the spindle) is most useful for finer threads. There are no spindle whorls known from Hornstaad Hörnle IA, which is not surprising as spindle whorls do not appear in the archaeological record of the Lake Constance area until the late Neolithic (*Endneolithischen*) Horgen culture (Schlichtherle 1990), although they are known in the Lagozza culture of northern Italy from the early fourth millennium (Baioni *et al.* 2003). Even when spindle whorls are present in later Pfyn sites, they appear in low numbers (Leuzinger 2002), which may suggest that other spinning methods were common. The tool used as a spindle for fine thread may simply be a stick rotated in the hand (Altorfer et Médard 2000, Leuzinger 2000) that would be difficult to recognise in archaeological context.

Spinning, time and place

Although there is little besides the threads themselves to investigate the time and place of thread production at Hornstaad Hörnle IA, what is apparent from historical or ethnographic accounts is that spinning is a time-consuming task. Many accounts talk of spinning and thread production carried out in short bursts alongside other tasks, filling all available hours of the day (Franquemont and Franquemont 1987). In other cases, large numbers of people are recruited to spin *en masse* to produce the required amount of thread (Flad 1984); either or both models may have been practised at Hornstaad Hörnle IA. The amount of time required to spin threads from different fibres and methods would have varied (fig. 4). When comparing the amount of time spent spinning

thread in relation to cloth construction, spinning thread is usually a more lengthy process than weaving (Evans 1985). An experienced spinner and weaver who was working on the construction of a Neolithic type plain-weave linen cloth for a museum reconstruction suggested that the spinning would take roughly ten times longer than the weaving (R. Sinkkonen-Davies pers. comm.).

Spinning with a spindle is a particularly mobile task: fibres and tools (spindle) can be carried and worked practically anywhere, and there are examples of people spinning when walking, standing and while engaged tasks such as watching livestock (Evans 1985, Franquemont and Franquemont 1987), although sitting is more usual (Flad 1984). From my own experiments with students, I found that the method of spinning by twisting on the thigh is most easily practised sitting. Photographs of this technique from ethnographic sources also show participants sitting (Mackenzie 1991, Médard 2003).

Constructing cloth

There are few tools associated with cloth construction at Hornstaad Hörnle IA, so the evidence for this stage of the process comes from the analysis of cloth artefacts, comparison with historical techniques and wider knowledge of cloth production in the Neolithic.

Netting

The nets are made of linen and can be divided into two types: wide-mesh nets made with fine two-plied threads (mostly less than 1 mm diameter) constructed with fishing net knots (*Filetknoten*) and coarser nets (mostly over 1 mm thread diameter) with smaller meshes made with lake-dwelling knots (*Pfahlbauknoten*) (Körber-Grohne und Feldtkeller 1998). Several examples have starting cords from tree bast. Through an analysis

Spinning method	Fibre	Time	Length	Source
Hand spinning by twisting on thigh	Tree bast fibres from *Ficus*	50-60 mins	10 m	20th century, Papua New Guinea (MacKenzie 1991)
Spinning with a spindle whorl	Flax	60 mins	250 m	19th century, Germany (Flad 1984)

Figure 4. Examples of thread production times: times required to produce threads using different spinning methods based on ethnographic and historical sources (Mackenzie 1991, Flad 1984).

of the way the knots are worked, Körber-Grohne and Feldtkeller suggest that the nets were probably knotted in the round, creating a tubular construction; only one net was worked in rows as if to create a flat net. In comparison with historical examples, it can be deduced that netting techniques usually begin from a row worked onto a taut starting cord or rod, with each subsequent row knotting onto the previous one (Geraint Jenkins 1974). According to Jenkins, historically equipment included a needle or bobbin to hold the thread and spacer to regulate the mesh width. Netting tools are not recognised at Hornstaad, therefore leaving no trace of when or where this task was carried out. In historical examples netting was readily carried out sitting in or near the home (op. cit.). Dieckmann's spatial analysis (1991) of nets throughout the village shows that many households possessed nets and net-sinkers.

Twining

There are many variations of twining at Hornstaad Hörnle IA as described by Körber-Grohne and Feldtkeller (1998). One noticeable group are the sieve bases. These lattices are made of fine tree bast threads and fixed into the base of a circular basket construction. Generally of thicker threads (still tree bast), there are a number of twined cloth fragments with close passive elements (vertical), but spaces left between each twined (horizontal) row. The twining technique is also worked with the rows close together to create a dense surface. Another distinct group are the twined cloth constructions with tufts added into one side, creating a fleecy effect. The size of some fragments, up to 33 x 51 cm and groups of fragments probably belonging to the same piece (Schlichtherle 1990) suggest some of these tufted twined cloths were originally large. Others are smaller, for example a tufted conical hat from tree bast measuring 16 x 20 cm (Feldtkeller und Schlichtherle 1987, Bazzanella et al. 2003b).

The villagers were clearly able to produce (or otherwise obtain) a wide range of cloth types using the twining technique. Depending on the size and flexibility of the threads and the size of the finished article, this could have been worked on a simple frame, suspended or lying on the ground, or worked by hand without a frame (Schlichtherle 1990, Feldtkeller 2004). With such variations, it is difficult to judge the amount of time to produce a piece of twined cloth. However, from my own experiments two novices working together twined a piece of cloth 30 x 23 cm with 3 cm spaces between the rows in 3 hours. This gives a rough idea that a large piece of cloth could be worked in days or weeks, rather than years.

Weaving

There are five documented small fragments of woven textiles in plain weave. They are from fine two plied linen threads. The threads are from 0,3-1 mm in diameter and are woven with from 6 to 10 threads per cm (Körber-Grohne und Feldtkeller op. cit.). In the Neolithic, woven textiles are associated with the warp-weighted loom. With no loom-weights found in the village, it is difficult to judge how the cloth was made. Other types of loom, such as a back-strap loom are a possibility (Bazzanella et al. 2003a), but difficult to prove here.

Working threads and cloth in the village and home

With no evidence of the frame or looms used to twine or weave cloth it is difficult to judge where and when these artefacts were made, although like the nets, a home or village location seems likely. The house size at Hornstaad Hornle IA, 3,5 x 9 m is big enough to set up a loom or other frame device, especially if it was vertical or resting against a wall. Neolithic sites in the later Neolithic show that warp-weighted looms were set up in lake-dwelling houses, in some cases, one in each house (Rast-Eicher 1992).

As each house at Hornstaad Hörnle IA appears to have had similar tool kits and house inventories, Dieckmann (1991) suggests that the members of each house were responsible for producing its own resources. This is further supported by the evidence for grain stores in each house, recognisable in the burnt layer, again suggesting that each household was responsible for its own needs (Maier 1999). As the villagers were preparing the flax and tree bast themselves, despite the lack of evidence for spinning and cloth construction tools, it seems highly likely that at least some of the thread and cloth were produced in the village; if not all of it. On this basis, it seems reasonable to assume that each household spun and made cloth for its own, varied requirements.

RESULTS

On the basis of the evidence outlined above, here I bring together the evidence for time and place in the tasks of processing plants for cloth. I do this through considering a " taskscape " of where and when the processing was carried out, in relation to the life-span of the villagers, as an aspect of planning resources and occasions and as a part of wider beliefs.

Constructing a " *taskscape* "

Digging and tending flax fields

At Hornstaad Hörnle IA, flax would have been sown in prepared fields either in spring (summer flax) or late autumn (winter flax) (fig. 5). In terms of place, flax cultivation would have been in the cultivated land area along with cereal crops, about 300-700 m from the village (fig. 6). The fields of flax appear to have been grown closer to the woodland than the cereal crops.

Digging the ground, sowing, weeding and care of flax crops would have required constant care; this would have meant individuals or small groups tending the fields on a regular basis, creating a repeated pattern of visiting the same place with seasonal variations in frequency and intent. A crop of flax could have been grown each year (fig. 7), either in the same field or following a crop rotation system and follows a similar seasonal pattern to cereals and poppies that were also cultivated.

	Jan	Feb	Mar	Apr	May	June	July	Aug	Sept	Oct	Nov	Dec
Spring crops			Prepare ground, sow spring crops		Tend crop, growing season		Harvest cereals flax and poppies					
Winter crops					Tend crop, growing season		Harvest cereals flax and poppies				Prepare ground, sow winter crops	
Spring collection			Collect tree bast									
Winter collection									Collect tree bast			
Wild fruits and berries gathering						Strawberries						
							Raspberries					
								Blackberries				
									Apples			
									Beech- and hazel-nuts			
										Sloes		

Figure 5. Seasonal cycles for cultivated and gathered crops at Hornstaad Hornle IA, showing alternative cycles for spring or autumn/winter flax cultivation and tree bast collection (based on a table from Maier 1990, p. 132, with additions).

Fibres from the forest

Lime, oak and elm trees grew in the forest area, possibly surrounding and beyond the fields (fig. 6). On the basis of comparisons with historically known practices, tree bast may have been collected in spring, when the bark is most easily removed from the tree, or in late autumn or winter, when the whole branch is heated to remove the bast (fig. 5). The method used to process tree bast at Hornstaad Hörnle IA was probably not water retting, so the heating method has been proposed. If collected in the autumn, this could have coincided with visits to the forest for wild fruits and berries. As each branch or trunk can only provide bast once: there must be a number of years of regrowth between each crop (fig. 7).

Therefore trees, or branches of a single tree, would be harvested in a rotating system over a number of years or decades. As a forest resource, it may have been loosely managed to prevent over-exploitation

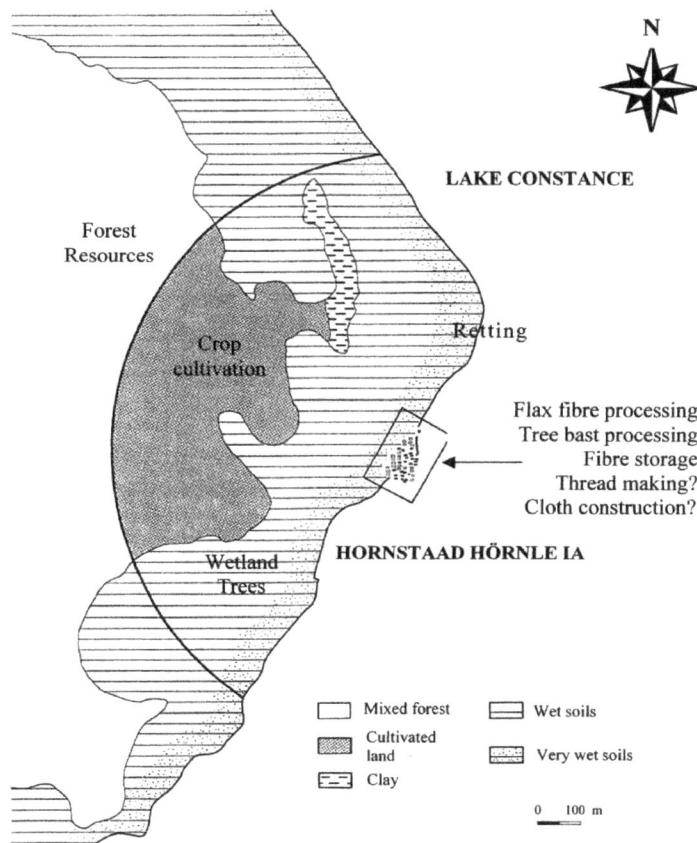

Figure 6. The place of cloth processing activities at Hornstaad Hörnle IA.
Tree bast collection from wetland trees or forest resources. Flax fields in crop cultivation area are closest to forest margins. Retting probably occurred in shallow water area. Flax plant storage, flax and tree bast processing carried out within the village. Spinning, weaving, twining and net construction probably occurred within the village (map from Maier and Vogt 2000, p. 125, annotations related to plant processing for cloth my own).

3917 cal. BC	3916	3915	3914	3913	3912	3911	3910	3909	3908	3907	3906	3905
Flax	Flax	Flax	Flax	Flax	Flax	Flax	Flax	Flax	Flax	Flax	Flax	Flax
Tree bast						Tree bast						
	Tree bast						Tree bast					

Figure 7. Cycles of annual flax growth, compared with a six-year time cycle for the growth of new branches or trees to harvest for bast.

of individual trees, or to encourage the growth of trees that they wanted to encourage; some form of coppicing may have occurred. Gathering tree bast would have required individuals or groups of people to walk through the forests to areas with suitable trees, in order to judge each branch for suitability in terms of species, age and quality. As lime, elm and oak grew in the vicinity, tree bast was readily available to the villagers.

Cloth in the village and home

Following the harvesting and gathering of flax or collection of tree bast, the dried stem bundles and bast strips may be processed the same year or be stored in the houses until there was time to prepare the fibres (fig. 8). Waste material from these processes suggests that this occurred throughout the village. In particular flax processing debris is present in all layers of the site and suggest that this task was also practised throughout the duration of the village.

The thread types present in the village show that the villagers were engaged in a number of different thread-producing activities, including plaiting, hand spinning by twisting on a surface such as the thigh, or spinning finer fibres with a spindle. As spinning is time-consuming and it appears that the villagers owned relatively large quantities of cloth items, spinning tasks were probably common and practised widely. Spinning was probably carried out in the village, but possibly also simultaneously with other tasks such as tending animals in the fields or forests.

Cloth construction for the household requirements probably occurred within the house or house area of the settlement, being attended to either on a daily or as-required basis. The construction of cloth such as nets or large sheets of twined or woven cloth would take up floor or wall space when being worked on. However, a net can be folded up during its construction and other cloth types may be left on the loom between work periods.

Process	Place	Time	Time duration
Flax fibre preparation	In concentrated areas around village	Annual, seasonal work?	Unspecified
Tree bast preparation	Unspecified, some debris in the village	Annual, seasonal work?	Unspecified
Spinning by twisting on the thigh	Sitting position	Regular, daily task or practised as required	Substantially longer than cloth construction time
Spinning with a spindle	Mobile, sitting, standing, walking		
Weaving or twining on a loom or frame	House? Loom or frame may be fixed in place, or movable between work sessions.		Finer and more dense constructions take longer than coarse, open constructions when using the same method
Twining without a loom	House? Worked from a sitting or standing position with no frame or loom.		
Netting	House? Worked from a starting line or frame, movable between work sessions.		

Figure 8. The places and time of cloth related tasks in the village.

Cloth processing during a lifetime

During this period, these processes must have repeated throughout the lives of the inhabitants. The village was established around 3917 cal BC and was abandoned some time after 3905 cal BC, young people would have grown up and adults grown older. Although there is little to indicate how men or women contributed to this *chaîne opératoire* at Hornstaad Hörnle IA, or more generally in the Neolithic, it is very likely that there was some gendered division of labour. Ethnographic surveys show the gender division of labour is strongly marked (Murdock 1967, Murdock and Provost 1973). In particular, it is common for craft industries to be exclusively worked by males or females and these may be kept distinct through physical and ideological boundaries (Costin 1996). Therefore, it is rare that men and women work in the same craft production, and when they do, they use different technologies, make different products or work for different purposes (*op. cit.*). While this might not make the actual gender attribution of these tasks in prehistory any clearer, it does suggest that the responsibility of the construction of co-existing cloth types, or different stages of the process may well have been the preserve of a particular gender group. Therefore, the development of skills throughout a man or woman's life could be related to both the age and gender role they were expected to fulfil.

A child in 3917 BC

So, for example, it would be likely that a child (boy or girl) who was 6 when the village was established was already being trained in appropriate skills such as spinning, digging, tending the fields, gathering in the forests and making cloth (fig. 9). These skills and practices would have developed over the years,

and young people would have been given increasing responsibility as appropriate to their gender role. At the time the village burned down, the young person would have been 13-14 years old, presumably with a number of personal responsibilities in some or all of the tasks involved in producing cloth depending on the way tasks were distributed in the village. The catastrophe of the village fire meant that valuable stores and seeds were lost, including fibres and crop seeds, as well as cloth artefacts such as fishing nets, clothing, tools and containers that were necessary for everyday life. These would have needed to be remade and there is evidence that production continued after the fire. This period probably increased the workload of all village members, as they rebuilt their lost homes and stores. Five years later, at 18, when the occupation of the village came to an end, the person may have had a family of his or her own to provide for; they may have been ready to leave the village taking their skills and cloth artefacts with them.

An adult in 3917 BC

In the second example, an adult man or woman aged 20 at the time the village was established would have already had many skills, as appropriate to their gender role. Some such individuals were surely involved with establishing fields and with the first crops of flax that were sown there, requiring a combination of tasks which were probably divided according to the gender of participants. During regular visits to the woods, these adults would have been familiar with the trees to use for tree bast and would have known where to go each year to find branches of a suitable age and quality; again the collection of such fibres may have been carried out by a particular gender group. At the time the village

3917 cal. BC	3916	3915	3914	3913	3912	3911	3910	3909	3908	3907	3906	3905
Establish fields and village	Village occupation						Burning incident/ rebuilding		Village occupation			
EXAMPLES OF AGE SEQUENCE												
First example starting at age 6, second example starting at age 20:												
Age 6	7	8	9	10	11	12	13	14	15	16	17	18
Age 20	21	22	23	24	25	26	27	28	29	30	31	32

Figure 9. Examples of the age of village inhabitants in relation to the duration of the village based on the stratigraphy and dendrochronology dates of building and burning phases.

burnt down, men and women in this age category were probably the most experienced and skilled village members, with responsibilities for younger members of the family, for coordinating the rebuilding of houses, and for the production of cloth equipment, passing on their skills of producing cloth to other members of the group. Besides this, men or women of this age may have been required to produce cloth for ceremonies such as marriages and burials or other rites-of-passage: they may have made cloth to exchange for other items, services or to establish social relationships.

The houses of the village are built close together in a community of several hundred people; men and women, boys and girls carrying out various tasks such as fibre processing, spinning, net-making or weaving. Cloth construction probably took up a sizeable amount of many individuals' time and was the source of social relationships and interaction. In the situation of the village, skills would be taught and passed on through the generations and between families. The continuity of cloth types at later sites in the Lake Constance area shows that skills were passed on through many generations.

Resources and planning ahead

Whether for clothing, to make fishing nets or other tools such as sieve bottoms or containers, people would have had to plan for their cloth needs. So, for example, if a plain weave flax cloth were needed for a particular event or purpose, the complete process of cloth production would need to be set in motion. Once in possession of a seed crop, the fields would need to be prepared and weeded, and the crop tended until harvest time. If the crop were successful, after harvesting it would need to be retted and dried, preparatory to processing. Processing was carried out in the village, at the earliest in the autumn after the harvest. At this point, nearly a year of work would have been invested in the flax. In addition, further time is necessary to spin and construct the cloth. However, this may have been spread out over a longer period of time as the prepared fibres, raw or retted flax stems and thread can be stored for years so long as they are protected from damp, fire and pests.

Without complete pieces of cloth, or a clear idea of the method of spinning and cloth construction, there is no accurate way of measuring the time taken to produce a piece of cloth, although this can be seen in a relative frame (fig. 8). Based on

my own experiments with twining cloth, this may have taken a few hours for a small piece such as a twined sieve bottom, or a few days for a piece of twined cloth with coarse thread and wide spacing. Larger pieces of other cloth types, for example a fine net, may have taken several days or months of work for an individual or small team working together.

Symbolism and beliefs

These combined passage of time and place, along with the social identity of the people creating the cloth, probably held their own social character in the village and accorded particular meanings and associated with symbolic events (prayers, festivals, ceremonies). For example, an event such as sowing a crop of flax or going to the forest to collect tree bast may have been quite a different social occasion to spinning or netting a fishing net: yet all are part of cloth processing. A particularly critical time may have been the sowing or harvesting of a crop, or the occasion of finishing a piece of cloth for an important event. Directing attention to these as aspects of time and place structures as contrasting parts of village life associated with cloth at Hornstaad Hörnle IA allows the possibility that they were valued and perceived in symbolic spheres, even though any value attached to these tasks remains unknown.

DISCUSSION

The aim of this paper was to take the evidence of plant processing to make cloth in the Neolithic into a realm that is tangibly human; the people who were engaged with making the artefacts were the focus of this paper, rather than just the processes or resulting artefacts. To do this took several stages: first was identifying a method.

Through reading about the role of cloth and technology in historical and ethnographic societies, I chose to investigate the time and place of each aspect of these processes. Time and place are socially significant to individuals and societies; they encompass the pattern of land use, the use of space in a village or home, they include the significance of seasonal rhythms, of people growing up and growing older, the " where " and " when " of people meeting and attending to one another. The place and time of such occasions are immediately social as a way people interact, but also are regularly imbued with symbolic or ideological significance, although the specifics of these may be elusive to archaeologists. The benefit of choosing such a method is that time and place are factors that can be addressed through the archaeological evidence: in terms of an excavation, through the location of finds, their presence in stratigraphic layers, the date of the site and the location of resources in and surrounding the site.

This method was then applied to a case study, Hornstaad Hörnle IA; a well-excavated and well-published Neolithic site with evidence for flax and tree bast cloth and cloth processing. The strength of this case study is that it is possible to trace time and place at a number of scales, for example the presence of plant processing and cloth in different layers of the stratigraphy and in relation to the dendrochronology dates of these layers, or the place of tasks in the houses, village or surrounding landscape. The weakness of this evidence is that some of the processes were not represented clearly at this site. For example, it is quite possible that fine thread could have been made with a simple stick that archaeologists would not recognise when excavating.

Not all time and place issues can be address through the excavation alone. A number of processes were identified through the post-excavation analysis of finds. Fortunately at this site, there are a number of detailed analyses of the botanical, pedological and cloth remains, which can be used to build up an idea of the time and place in the way people processed plants from the acquisition of raw materials to the construction of cloth. So for example, the botanical and pedo-logical analyses were useful in locating the area that was most probably used for cultivation. In other cases the post-excavation analyses were useful in recognising the relationship between the archaeological artefacts and experimental analyses or ethnographical and historical practices of these processes. This is especially the case with U. Körber-Grohne and A. Feldtkeller (1998) who identified that the tree bast was probably not processed by water retting at this site. They proposed alternative processes, with implications for the time and place of these practices.

However, not all approaches to the evidence are so satisfactory. The weakness of approaching the evidence through comparison with historical and ethnographic sources, or by experimental archaeology is that the results can only be used as a suggestion. So there are examples of the amount of time it takes to spin a length of thread with a spindle, or through spinning by twisting on the thigh, but these can only account for a realm of possibility, rather than establish what happened at Hornstaad Hörnle IA. Similarly, the place people spun such thread or made nets or wove cloth is not clear at this site, although there are a number of possibilities. Spinning with a spindle could have been a mobile or sedentary task, and on the basis of experimental experience, the houses at Hornstaad Hörnle IA were big enough to fit a warp-weighted loom.

These different sources of evidence have their own strengths and weaknesses, but overall, there is a benefit in trying to piece together the maximum amount of information on each of these tasks. This helps to create a wider appreciation of social issues such as planning resources, or the development and practice of a variety of skills throughout a lifetime. Through this it is possible to gain an understanding of the frequency the people repeated tasks on a daily, seasonal or annual basis, the changing tasks associated with different seasons, or particular places.

CONCLUSION

In the case study of the production of cloth from flax and tree bast at Hornstaad Hörnle IA, I have approached the evidence for plant processing through concepts of time and place within the village situation. I have considered people visiting places in the surrounding landscape to collect or cultivate the raw materials for cloth, and outlined the location and timing of people working regularly in fields planting annual crops with the exploitation of resources of tree bast from species of trees that needed years to rejuvenate and had to be harvested in cycles.

By comparison with other environmental evidence, these activities can be understood in relation to the season and location of other plant processing tasks, such as growing cereals or collecting wild apples or hazelnuts. After this, it seems that the processing and construction of the cloth was carried out in or near the village: the debris of flax fibre extraction or tree bast waste fell from house platforms to the water or ground below. Thread production and cloth construction are less clearly placed as there are few indications of how the villagers carried these out. Yet it seems likely that this did occur in or around the village, probably at a household level, not excluding that some items may have come from outside the area. The villagers probably spent more time spinning or plaiting fibres for threads, than cloth construction, whether weaving, twining or netting. But exactly how long these tasks took, or when they were performed in the day or year remains speculative.

Taking advantage of the narrow date range of this lake dwelling village it is possible to envisage these annual cycles and daily tasks in terms of single life-spans. In this way cloth production and use can be viewed in terms of skills being developed and passed on throughout men and women's lives and put into practice to replace cloth lost in the village fire. Consequently, cloth occupies a time frame of growing up and growing old and fulfilling the expectations of being a man or woman in the village.

In this way, through cloth, the inhabitants of Hornstaad Hörnle IA were negotiating the relationships necessary to produce cloth (shared tasks, teaching and learning), planning ahead to anticipate their needs (social and material) or evaluating what they had in relation to their needs or the judgement of others (material possessions). This occurred in a regular rhythm of annual cycles, of daily events that created the predictable shape of an individual's life (gender, age, abilities), periodical events which may have included marriage or funerary ceremonies and irregular events, such as the fire catastrophe and recovery from it.

This research is based on evidence from archaeological excavations, post-excavation analysis of artefacts, comparison of archaeological evidence with historical and ethnographic sources, and comparison with processes carried out through experimental archaeology. As noted in the discussion, the strength in using this range of evidence is that it allows a wide ranging investigation of issues of task, time and place. The weakness is that some lines of evidence are open to debate and there are several alternative possibilities. I hope that some of these points will become clearer through further study, especially through more results from experimental archaeology in relation to archaeological materials.

ACKNOWLEDGEMENTS

This research is based on work undertaken for my PhD at the Institute of Archaeology, University College London supervised by Prof. Ruth Whitehouse and Dr. Sue Hamilton. I am indebted for the Arts and Humanities Research Council (AHRC) for funding my PhD and the Institute of Archaeology UCL for funding a fieldwork visit to the research centre at Hemmenhofen, Germany to study the artefacts from Hornstaad Hörnle IA. Many thanks to Dr. Bodo Dieckmann and Dr. Hermut Schilchterle for their permission to view the Hornstaad Hörnle IA finds and to all the researchers at Hemmenhofen for their generous discussion of the site. Particular thanks to Dr. Ursula Maier and Richard Vogt for their kind permission to use their map for figure 6.

AUTHOR'S ADRESS

Susanna HARRIS
62 Brooke Road
Stoke Newington
London N16 7RU
Great Britain
susannaharris@hotmail.com

REFERENCES

ADAMS, M. J., 1971. Work Patterns and Symbolic Structures in a Village Culture, East Sumba, Indonesia. *Southeast Asia, An International Quarterly*, 1 (4), 321-334.

ALTORFER, K. et MÉDARD, F., 2000. Nouvelles découvertes textiles sur le site de Wetzikon-Robenhausen (Zürich, Suisse). Sondages 1999. *In :* D. Cardon et M. Feugère, eds. *Archéologie des textiles des origines au Vᵉ siècle, Actes du colloque de Lattes, octobre 1999.* Montagnac : Éditions Monique Mergoil, 35-75.

BAIONI, M., BORRELLO, M. A., FELDTKELLER, A., SCHLICHTHERLE, H., 2003. I pesi reniformi e le fusaiole piatte decorate della Cultura della Lagozza. Cronologia, distribuzione geografica e sperimentazioni. *In :* M. Bazzanella *et al.,* eds. *Textiles : intrecci e tessuti dalla preistoria europea.* Trento : Provincia Autonoma di Trento, Servizio Beni Culturali, Ufficio Beni Archeologici, 99-109.

BAZZANELLA, M., BELLI, R., MAYR, A., 2003a. Analisi sperimentali condotte sulla fascia decorata della palafitta di Molina di Ledro. *In :* P. Bellintani e L. Moser, eds. *Archeologie sperimentali. Metodologie ed esperienze fra verifica, riproduzioni, comunicazioine e simulazione. Atti del Convegno Comano Terme - Fiavè (Trento, Italy) 13-15 settembre 2001.* Trento : Provincia autonoma di Trento, Servizio beni culturali, Ufficio beni archeologici, 273-282.

BAZZANELLA, M., MAYR, A., MOSER, L., RAST-EICHER, A., 2003b. Schede [Catalogue]. *In* : M. Bazzanella *et al.,* eds. *Textiles: intrecci e tessuti dalla preistoria europea.* Trento : Provincia Autonoma di Trento, Servizio Beni Culturali, 133-289.

BILLAMBOZ, A., 1990. Der Holz der Pfahlbausiedlungen Südwestdeutschlands: Jahrringanalyse aus archäodendrologischer Sicht. *Bericht der Römisch-Germanischen Kommission,* 71, 187-206.

BILLAMBOZ, A., 1998. Die Jungneolithischen Dendrodaten der Pfahlbausiedlungen Südwestdeutschlands als Zeitrahmen für die Einflüsse der Michelsberger Kultur in ihrem südlichen Randgebeit. *In*: J. Biel, ed. *Die Michelsberger Kultur und ihre Randgebiete. Probleme der Entstehung, Chronologie und des Siedungswesens.* Stuttgart: Materialh. Arch. Baden-Württemberg 43, 159.

COSTIN, C. L., 1996. Exploring the Relationship Between Gender and Craft in Complex Societies: Methodological and Theoretical Issues of Gender Attribution. *In*: R. Wright, ed. *Gender and Archaeology.* Philadelphia: University of Pennsylvania, 111-142.

DIECKMANN, B., 1987. Ein bemerkenswerter Kupferfund aus der jungsteinzeitlichen Seeufersiedlung Hornstaad-Hörnle I am westlichen Bodensee. *Archäologische Nachrichten aus Baden,* 38/39, 28-37.

DIECKMANN, B., 1991. Zum Stand der archäologischen Untersuchungen in Hornstaad. *Bericht der Römisch-Germanischen Kommission,* 71, 84-109.

DOBRES, M.-A., 2000. *Technology and Social Agency: Outlining a Practice Framework for Archaeology.* Oxford: Blackwell.

EVANS, N., 1985. *The East Anglian Linen Industry: Rural Industry and Local Economy 1500-1850.* London: Gower, The Pasold Research Fund.

FELDTKELLER, A., 2004. Die Textilen von Seekirch-Achwiesen. *In*: *Ökonomischer und ökologischer Wandel am vorgeschichtlichen Federsee.* Landesdenkmalamt Baden-Württemberg, Gaienhofen-Hemmenhofen, 56-70.

FELDTKELLER, A. und SCHLICHTHERLE, H., 1987. Jungsteinzeitliche Kleidungsstücke aus Ufersiedlungen des Bodensees. *Archäologische Nachrichten aus Baden,* 38/39, 74-84.

FLAD, M., 1984. *Flachs und Leinen: Vom Flachsanbau, Spinnen und Weben in Oberschwaben und auf der Alb.* Ravensburg: Schwäbischer Bauer.

FRANQUEMONT, E. and FRANQUEMONT, C., 1987. Learning to Weave in Chinchero. *The Textile Museum Journal,* 26, 55-79.

GERAINT JENKINS, J., 1974. *Nets and Coracles.* London: Newton Abbot.

HOFFSTADT, J. und MAIER, U., 1999. Handelsbeziehungen während des Jungneolithikums im Westlichen Bodenseeruam am Beispiel der Fundplätze Mooshof und Hornstaad Hörnle IA. *Archäologisches Korrespondenzblatt,* 29, 21-34.

INGOLD, T., 1993. The temporality of the landscape. *World Archaeology,* 25 (2), 152-174.

KOKABI, M., 1990. Ergebnisse der osteologischen Untersuchungen an den Knochenfunden von Hornstaad im Vergleich zu anderen Feuchtbodenfundkomplexen Südwestdeutschlands. *Bericht der Römisch-Germanischen Kommission,* 71, 145-160.

KÖRBER-GROHNE, U. und FELDTKELLER, A., 1998. Pflanzliche Rohmaterialien und Herstellungstechniken der Gewebe, Netze, Geflechte sowie anderer Produkte aus den neolithischen Siedlungen Hornstaad, Wangen, Allensbach und Sipplingen am Bodensee. *In* : *Siedlungsarchäologie im Alpenvorland V.* Stuttgart : Konrad Theiss Verlag, 131-189.

LEMONNIER, P., 1993. Introduction. *In*: P. Lemonnier, ed. *Technological Choices: Transformation in Material Cultures since the Neolithic.* London: Routledge, 1-35.

LEUZINGER, U., 2002. Textilherstellung. In : A. de Capitani *et al.,* eds. *Die jungsteinzeitliche Seeufersiedlung Arbon-Bleiche 3 : Funde.* Frauenfeld : Archäologie im Thurgau, 11, 115-134.

MACKENZIE, M. A., 1991. *Androgynous objects: string bags and gender in central New Guinea.* Chur: Harwood Academic Publishers.

MAIER, U., 1990. Botanische Untersuchung in Hornstaad-Hörnle IA. Neue Ergebnisse zu Landwirtschagt und Ernährung einer jungsteinzeitlichen Uferrandsiedlung. *Bericht der Römisch-Germanischen Kommission,* 71, 110-135.

MAIER, U., 1999. Agricultural activities and land use in a Neolithic village around 3900 BC: Hornstaad Hörnle IA, Lake Constance, Germany. *Vegetation History and Archaeobotany,* 8, 87-94.

MAIER, U., 2001. Archäobotanische Untersuchungen in der neolithischen Ufersiedlung Hornstaad Hörnle IA am Bodensee. *In* : U. Maier and R. Vogt, ed. *Siedlungsarchäologie im Alpenvorland VI /Botanische und pedologische Untersuchungen zur Ufersiedlung Hornstaad-Hörnle IA, vol. 74.* Stuttgart : Konrad Theiss, 9-384.

MAIER, U. and VOGT, R., 2000. Reconstructing the neolithic landscape at Western Lake Constance, Germany. *Archaeology in the Severn Estuary,* 11, 121-130.

MÉDARD, F., 2000. *L'artisanat textile au Néolithique. L'exemple du Delley-Portalban II (Suisse), 3272-2462 avant J.-C.* Montagnac : Éditions Monique Mergoil.

MÉDARD, F., 2003. La produzione di filo nei siti lacustri del Neolitico. In : M. Bazzanella *et al.*, eds. *Textiles: intrecci e tessuti dalla preistoria europea.* Trento : Provincia Autonoma di Trento, Servizio Beni Culturali, Ufficio Beni Archeologici, 79-86.

MOSER, F., et MEDARD, F., 2001. Quelques observations concernant la fabrication expérimentale des étoffes cordées. *Annales des IXe Rencontres Archéologiques de Saint-Céré (Lot)*, 8, 63-73

MÜLLER, A., 1994. Geflecht und Gewebe aus Hornstaad Hörnle I. *In*: G. Jaacks, and K. Tidow, eds. *North European Symposium for Archaeological Textiles (5th, 1993, Neumünster, Germany) : Textilsymposium Neumünster.* Neumünster : Textilmuseum Neumünster, 27-33.

MURDOCK, G., 1967. *Ethnographic Atlas.* Pittsburgh: University of Pittsburgh Press.

MURDOCK, G. and PROVOST, C., 1973. Factors in the Division of Labor by Sex: A Cross Cultural Analysis, *Ethnology*, 12 (203), 225.

MYKING, T., HERTZBERG, A. and SKRØPPA, T., 2005. History, manufacture and properties of lime bast cordage in Northern Europe. *Forestry*, 78 (1), 65-71.

PFAFFENBERGER, B., 1988. Fetishised Objects and Humanised Nature: Towards an Anthropology of Technology. *Man*, 23, 236-252.

RAST-EICHER, A., 1992. Die Entwicklung der Webstühle vom Neolithikum bis zum Mittelalter. *Helvetia Archaeologica*, 23, 56-70.

RAST-EICHER, A., 2005. Bast before Wool: the first textiles. *In*: P. Bichler *et al.*, eds. *Hallstatt Textiles: Technical Analysis, Scientific Investigation and Experiments on Iron Age Textiles.* BAR International Serie 1351, Oxford: Archaeopress, 117-131.

RÖSCH, M., 1996. New approaches to prehistoric land-use reconstruction in south-western Germany. *Vegetation History and Archaeobotany*, 5 (65), 79.

RÖSCH, M., EHRMANN, O., HERRMANN, L., SCHULZ, E., BOGENRIEDER, A., GOLDAMMER, J. P., HALL, M., PAGE, H. and SCHIER, W., 2002. An experimental approach to Neolithic shifting cultivation. *Vegetation History and Archaeobotany*, 11, 143-154.

SCHLICHTHERLE, H., 1990. *Siedlungsarchäologie im Alpenvorland / Landesdenkmalamt Baden-Württemberg I : Die Sondagen 1973-1978 in den Ufersiedlung Hornstaad-Hörnle I.* Stuttgart : Konrad Theiss Verlag.

VOGT, R., 1990. Pedologische Untersuchungen im Umfeld der neolithischen Ufersiedlungen Hornstaad-Hörnle. *Bericht der Römisch-Germanischen Kommission*, 71b (1), 136-144.

AROUMANS
(*ISCHNOSIPHON* SPP., MARANTACEAE).
VANNERIE ET SYMBOLISME EN GUYANE FRANÇAISE

Damien DAVY

Résumé : Cette étude est consacrée à la vannerie chez les populations amérindiennes, créoles et noirs marrons de Guyane française. Elle est fondée sur une enquête ethnographique et ethnobotanique menée, entre 2002 et 2006, auprès de 18 communautés et 168 artisans vanniers, établis sur le littoral et à l'intérieur de ce département, appartenant aux groupes créoles, noirs marrons aluku, arawak-lokono, kali'na, palikur, teko, wayana et wayãpi. Dans le cadre de cet article, on s'attachera tout particulièrement à décrire la chaîne opératoire de fabrication des vanneries et à définir la place qu'occupe cet artisanat, tant au niveau social, culturel que symbolique. En Guyane française, la vannerie tient notamment une place essentielle dans la transformation du manioc amer, base de l'alimentation de ces sociétés. Elle est aussi l'objet de nombreux rites et interdits et se trouve au cœur de nombreux mythes. Elle est aussi un marqueur ethnique important.

Mots-clés : *Ischnosiphon* spp., vannerie, Guyane française, ethnobotanie, symbolisme.

Abstract: This study deals with basketry of Amerindians, Creoles and Bush Negroes of French Guiana. It is based on ethnographic and ethnobotanic data collected between 2002 and 2006 amongst 18 communities and 168 craftsmen living on the coast and inland, belonging to the Creole, Noirs Marrons Aluku, Arawak-Lokono, Kali'na, Palikur, Teko, Wayana et Wayãpi. The present paper mainly focuses on the basketry manufacturing and the role of this handicraft from a technical, cultural and symbolic point of view. In French Guiana basketry is important in the transformation process of the bitter cassava, staple diet of these societies. Basketry processing is also linked with ritual practises, taboos and myths. Furthermore it plays a role as ethnic marker.

Keywords: *Ischnosiphon* spp., basketry, French Guiana, ethnobotany, symbolism.

INTRODUCTION

Les sociétés forestières amérindiennes d'Amérique du sud sont encore des sociétés tirant une part majeure de leurs ressources du monde végétal. Outre les plantes médicinales, rituelles et alimentaires, la forêt guyanaise est aussi source de nombreux végétaux à usages techniques (c'est-à-dire utiles pour l'artisanat, la construction …), ce que certains nomment une véritable « forêt matière » (Bahuchet 2000). Avec 5000 espèces végétales supérieures estimées, dont 1200 espèces de ligneux, cette région est depuis longtemps un réservoir pour de nombreuses plantes utiles aux différentes communautés y vivant.

Si les sociétés amérindiennes d'Amazonie et des Guyanes subissent de profonds bouleversements dans leur mode de vie depuis plusieurs siècles, avec une accélération certaine depuis quelques décennies, elles conservent néanmoins encore une connaissance fine de leur environnement.

La Guyane française, située au nord-est du continent sud-américain, département français et morceau d'Europe en Amazonie (fig. 1), est à bien des égards un cas particulier en Amérique du Sud puisque les changements culturels y sont surtout liés à l'assistanat.

Il se dégage de notre étude que la pratique de la vannerie reste encore assez active dans cette région, même si les artisans vieillissent et que l'utilisation domestique de ces vanneries recule au profit d'une commercialisation grandissante. Néanmoins, la vannerie reste l'activité artisanale la plus importante chez les Amérindiens de Guyane.

Depuis quatre ans, dans le cadre d'une thèse de doctorat, nous menons une étude sur cet artisanat en essayant d'appréhender le changement que connaît cette activité au contact de notre société. Les résultats présentés ici sont issus de vingt mois de terrain.

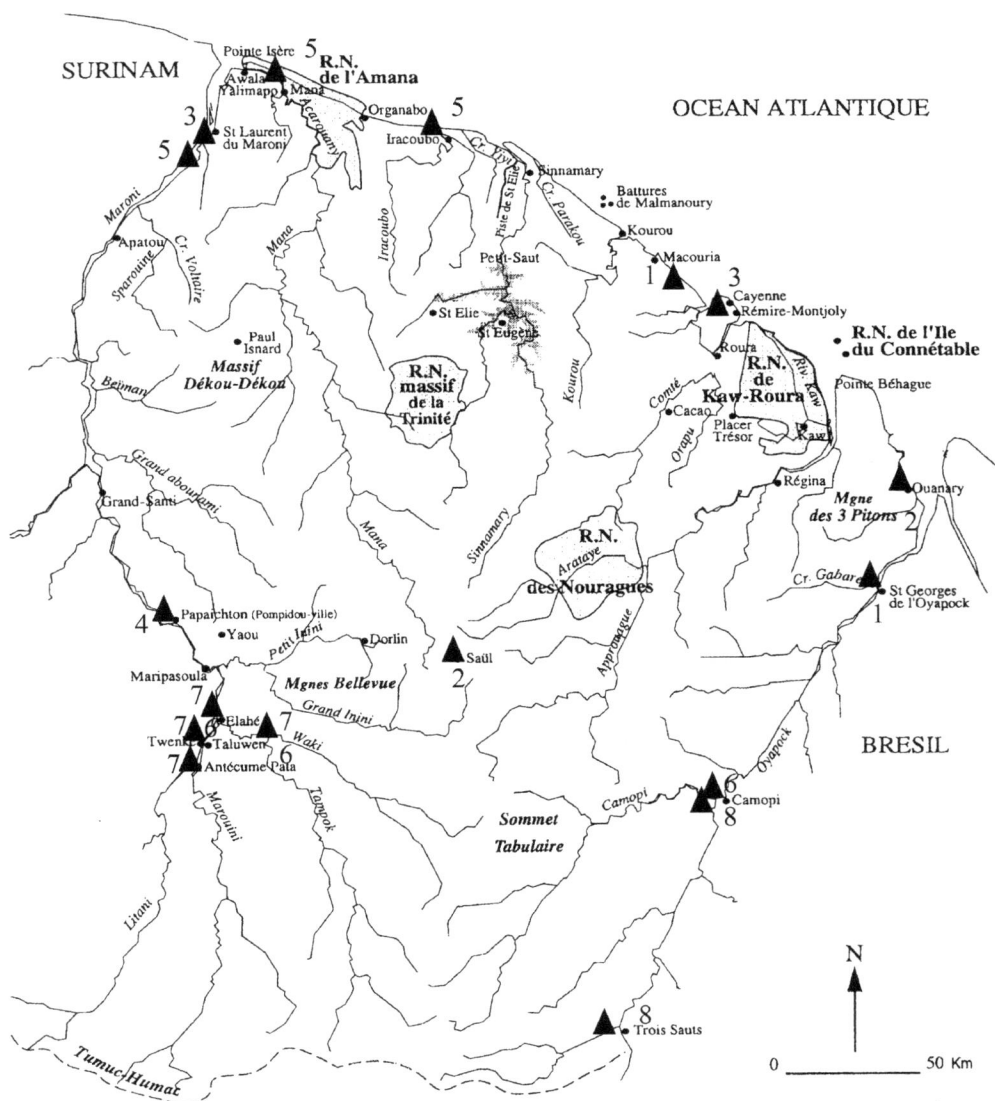

Figure 1. Carte de la Guyane française et lieux d'étude (▲). (1) population palikur, (2) population créole, (3) population arawak, (4) population aluku, (5) population kali'na, (6) population teko, (7) population wayana-apalai, (8) population wayãpi.

Dans l'article qui suit, nous présenterons plus particulièrement la place que tient, chez ces peuples, la vannerie tant d'un point de vue alimentaire, identitaire que symbolique. Cette étude est fondée sur une enquête ethnographique et ethnobotanique menée, entre 2002 et 2006, auprès de 18 communautés et 168 artisans vanniers, établis en Guyane française, sur le littoral et à l'intérieur du département, appartenant aux groupes créoles[1], noirs marrons[2] aluku et arawak-lokono, kali'na, palikur, teko, wayana et wayãpi[3].

La Guyane étant couverte à 90 % par la forêt dense humide, ces habitants ont appris à en vivre et tirer le meilleur de cet environnement riche et varié. Les communautés étudiées sont principalement rurales, même si celles de l'intérieur du territoire (les Teko,

[1.] Les populations créoles de Guyane sont les descendants, souvent métissés, des esclaves libérés à l'abolition de l'esclavage ayant eu lieu en 1848. Ils représentent 40 % de la population et sont donc la population la plus nombreuse en Guyane. Ils vivent principalement en ville même s'il reste encore quelques petits bourgs ruraux créoles.

[2.] Les Noirs Marrons sont les descendants d'esclaves africains ayant fui les plantations au cours du XVIIIe siècle. Un groupe nommé aluku vit entièrement sur le territoire français tandis que trois autres, les Ndjuka, les Saramaka et les Paramaka, habitent principalement au Surinam mais sont également très présents en Guyane française.

[3.] La Guyane française est habitée par six ethnies amérindiennes comptabilisant au total environ 6500 personnes : les Arawak-Lokono et les Palikur du groupe linguistique arawak, les Kali'na et les Wayana-Apalai du groupe linguistique karib, enfin les Teko et les Wayãpi du groupe tupi-guarani.

les Wayana, les Wayãpi, les Aluku vivant sur les fleuves Maroni et Oyapock) ainsi que les Créoles de Saül et surtout de Ouanary sont beaucoup plus isolés que les Kali'na, les Arawak-Lokono et les Palikur, habitant le littoral jamais très loin de bourgs urbains. Si les communautés les plus isolées comme celles du haut et moyen Oyapock ainsi que du haut Maroni vivent dans des villages équipés de centres de santé et d'eau courante, une grande partie de leur ressource provient toujours de l'agriculture (culture sur brûlis ou abattis), de la chasse et de la pêche. Ceux ayant la nationalité française perçoivent, en plus, des aides sociales qui leur permettent d'acheter essence, moteur, fusil et autres denrées occidentales dont ils sont de plus en plus dépendants. Pour ces populations de l'intérieur, à part pour les Aluku, les travaux salariés sont rares et pour la plupart attachés à la fonction publique (auxiliaires de santé, piroguiers de l'armée ou de la gendarmerie, agents communaux …). Les populations du littoral sont elles beaucoup plus dépendantes du monde occidental, le travail salarié, légal ou non, étant plus important même si elles pratiquent encore l'agriculture, la chasse et la pêche.

Les vanneries, faisant l'objet de cet article, étaient toutes, à l'origine, tressées pour un usage domestique. Aujourd'hui, que ce soit sur le littoral ou dans l'intérieur, cette activité est en plein bouleversement (Davy 2002, 2006). Cependant les hottes, tamis, presses à manioc, éventails à feu et autres divers paniers (figs 2a-e) sont encore régulièrement tressés et utilisés et ce principalement avec la plante phare de cette activité : l'arouman.

Figure 2. Vanneries de Guyane française (ci-dessus et pages suivantes) (photos de l'auteur).
a. *Yamat:u* ou coffret kali'na en arouman, avec motif *akuwamay* symbolisant un être mythique.

b. *Yamat:u* ou coffre kali'na pour ranger le linge,
les instruments des chamanes ainsi que les plumasseries avec motif *oruki* (chenille).

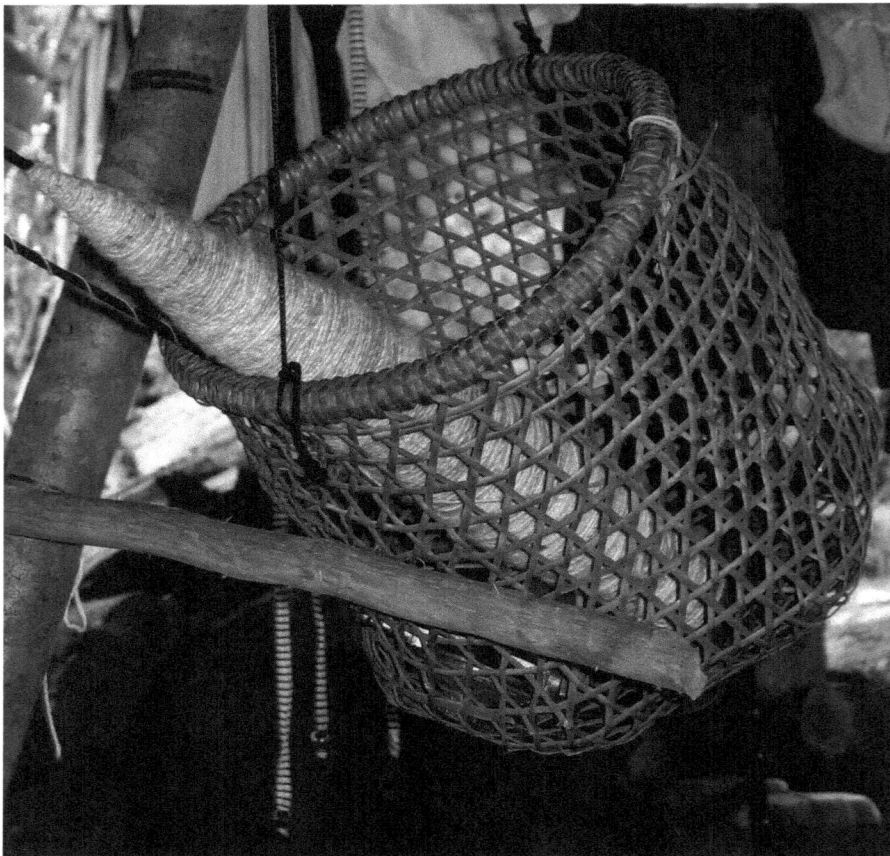

c. Panier ajouré wayana en liane, nommé *pïlasi piya ewu* (panier œil d'aigle harpie).

d. Panier à pied wayana avec motif *tsikalewöt* (chenille mythique poilue).

e. Tamis à manioc et corbeille wayãpi pour recueillir la farine tamisée.
La corbeille ou *panakali* est décorée du motif *iwïtaolape* (chemin de l'amphisbène).

LA CHAÎNE OPÉRATOIRE DE FABRICATION DES VANNERIES

La matière première : l'arouman, une plante à vannerie emblématique

En Guyane française, région d'une très grande biodiversité (Blanc *et al.* 2003), les sociétés forestières disposent d'une grande quantité d'espèces végétales. Pour la seule activité de vannerie, nous avons recensé 125 espèces différentes de végétaux utilisés par tous les groupes étudiés, c'est-à-dire les six ethnies amérindiennes, les Créoles et les Noirs Marrons.

Ces 125 espèces végétales appartiennent à 38 familles botaniques différentes. Dans ce total, sont comptabilisées les plantes utilisées dans la vannerie au sens large, c'est-à-dire qu'en plus des plantes que l'on tresse, se trouvent également :

- Les plantes utilisées pour leur bois et destinées à consolider les bords des vanneries et à faire des armatures et des pieds.

- Différentes écorces servant à confectionner les bretelles des hottes.

- Les plantes donnant des fibres, des teintures naturelles et des résines.

- Les feuilles nécessaires pour imperméabiliser les coffrets et les chapeaux.

Toutes les parties des végétaux s'utilisent donc : tige, branche, fruit, résine, feuille, racine aérienne et écorce.

Rares sont les vanneries ne comportant comme matière première qu'une espèce végétale. La plupart sont de véritables œuvres composites nécessitant deux, trois, voire sept espèces pour les ouvrages les plus élaborés. Le tableau (fig. 3) recense les dix principales familles botaniques avec, pour chacune d'entre elles, les parties des plantes utilisées et leurs usages.

On peut remarquer que la famille des palmiers (Arecaceae) offre le plus grand nombre d'espèces à tresser (19). Mais, comme nous pouvons l'observer dans la dernière colonne, cette famille n'est utilisée que pour 25 % des formes tressées. Par contre, 70 % des formes sont confectionnées avec des plantes d'une famille plus faiblement représentée en nombre d'espèces, les Marantacées. Cette famille comporte des espèces majeures pour la vannerie de Guyane et plus largement du plateau des Guyanes et du nord-ouest de l'Amazonie, les aroumans. Il a été montré (Ribeiro 1985) que la vannerie de cette région reste dominée par les Marantacées, alors que le sud et le centre de l'Amazonie paraissent dominés par l'usage des palmiers. L'arouman est bien la plante emblématique de la vannerie du plateau des Guyanes.

Les aroumans du genre *Ischnosiphon* regroupent essentiellement deux espèces largement utilisées en Guyane ainsi que deux autres moins prisées par les artisans. La sélection se fait suivant des critères techniques ainsi que suivant leur disponibilité, le facteur le plus important étant les caractères techniques. Les deux espèces les moins prisées (*I. puberulus* et *I. centricifolius*) n'offrent pas un

Familles	Nombre d'espèces	Usages principaux	Principale partie utilisée	% de formes tressées par famille
Arecaceae	**19**	**Tressage de vannerie**	**pousse foliaire**	**25%**
Lecythidaceae	14	Bretelles de hotte	écorce	-
Annonaceae	8	Bretelles de hotte	écorce	-
Myrtaceae	8	Baguettes pour armatures	bois	-
Chrysobalanaceae	6	Baguettes pour armatures	bois	-
Melastomataceae	5	Baguettes pour armatures	bois	-
Burseraceae	4	Teinture	résine	-
Marantaceae	**4**	**Tressage de vannerie**	**tige**	**70%**
Araceae	**4**	**Tressage de vannerie**	**racine aérienne**	**5%**
Mimosaceae	3	Fixateur pour teinture	résine	-
Bixaceae	1	Teinture	fruit	-

Figure 3. Principales familles et parties de plantes utilisées.

brin assez long en raison de leur port lianescent et de leurs courts entre-nœuds[4]. Elles possèdent, de plus, une tige plus dure et plus rigide entraînant ainsi une moins bonne qualité de fibre.

Les deux espèces les plus utilisées sont *Ischnosiphon arouma* et *Ischnosiphon obliquus*, respectivement l'*arouman rouge* et l'*arouman blanc* en créole (noms donnés en raison de la teinte que prend le brin au bout d'un certain temps lorsqu'il est tressé non gratté) (fig. 4).

Ce sont des arbustes constitués d'une longue tige chlorophyllienne terminée par un nœud et un bouquet foliaire comportant également les inflorescences. Un individu est constitué de plusieurs tiges partant de la même racine. Cette plante peut atteindre quatre mètres.

L'*arouman blanc* pousse préférentiellement dans les bas-fonds humides et est souvent présent en

---

[4.] En botanique, un entre-nœud est la partie d'une tige comprise entre deux nœuds foliaires ou floraux.

grande quantité dans un même peuplement végétal tandis que l'*arouman rouge* se trouve plus disséminé et pousse plutôt en terrain drainé. Comme la quasi-totalité des espèces utilisées dans la vannerie, ces plantes sont cueillies à l'état sauvage dans la forêt ancienne. Les plants d'arouman poussant dans les zones secondarisées, c'est-à-dire qui ont été défrichées pour l'agriculture dans le passé comme les anciennes cultures sur brûlis, ne sont pas très prisés en raison de leurs moindres qualités techniques.

De la plante au brin

Nous allons détailler ci-après la séquence de gestes et d'actions nécessaire pour transformer une tige d'arouman en un brin utilisable (fig. 5). Précisons que la préparation des brins est un travail masculin comme l'activité de vannerie en général, mais nous y reviendrons. Quand un artisan commence à préparer les brins, il essaye de terminer cette séquence dans la journée afin de ne pas laisser les tiges trop sécher au risque de les rendre inutilisables. Les artisans préparent toutes les tiges utiles, en même temps, étape par étape :

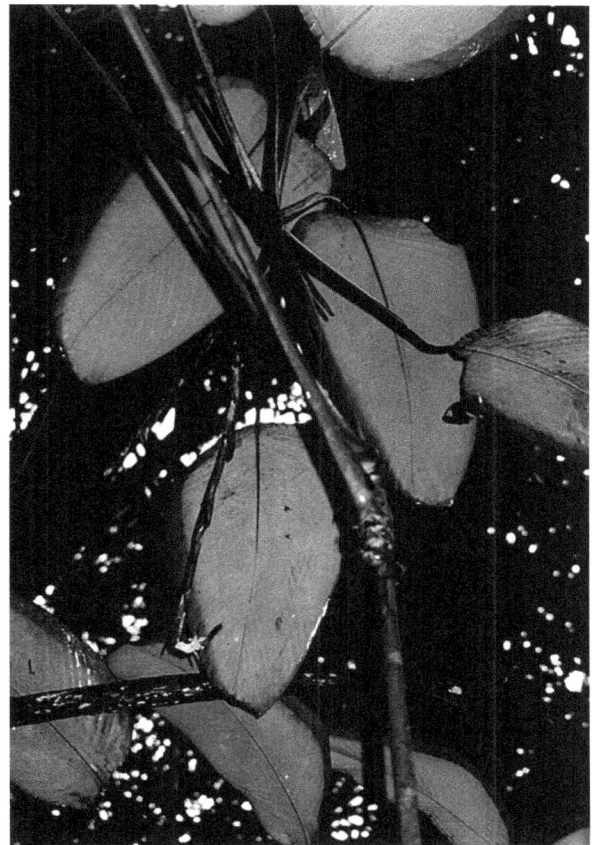

Figure 4. À droite, bouquet d'*Ischnosiphon arouma* avec inflorescence et tige, à gauche, d'*Ischnosiphon obliquus* (photos de l'auteur).

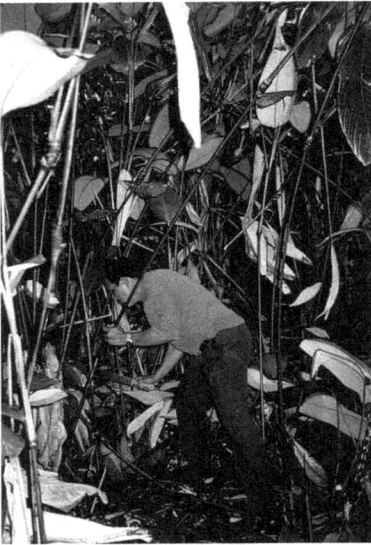

a

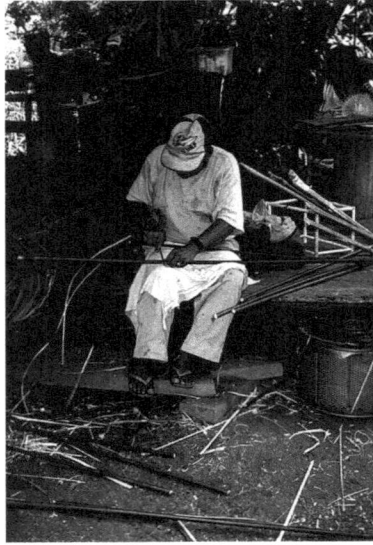

b

c

d

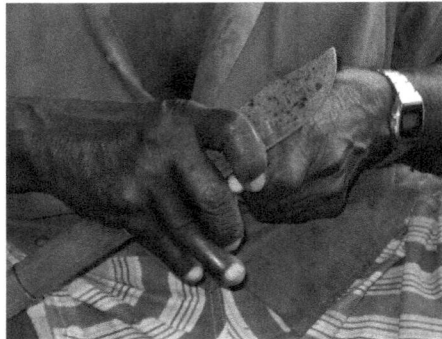

e

f

g

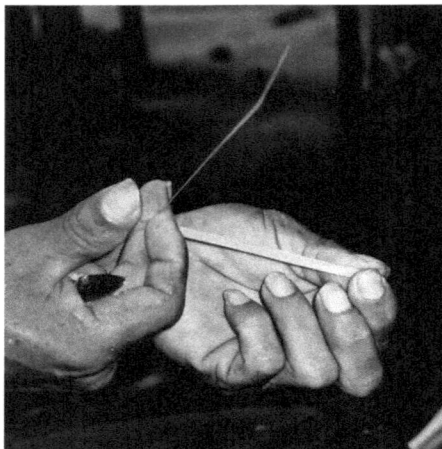

h

j

Figure 5. De la tige au brin : chaîne opératoire de la transformation de l'arouman (*Ischnosiphon* spp.).
Chaque lettre en bas à droite des photos renvoie aux étapes énumérées dans le texte (photos de l'auteur).

a. Récolter

Les aroumans sont récoltés par les hommes lors de sorties spécifiques. Les zones sont, soit découvertes lors de sorties de chasse, soit ce sont des zones dans lesquelles l'artisan à l'habitude de se rendre. En effet, bien souvent chaque artisan connaît une ou plusieurs zones à arouman dans lesquelles il va puiser régulièrement. Il utilise la machette pour la récolte. À chaque sortie, il ne récolte que le nombre de tiges dont il a besoin. En effet, une fois coupées, les tiges doivent être travaillées assez vite sous peine de se dessécher. Si elles ne sont pas travaillées immédiatement, elles sont conservées à l'ombre ou enfouies dans le sable ou dans l'eau d'un cours d'eau.

b. Couper

Les tiges sont coupées de la longueur estimée nécessaire en fonction de l'ouvrage à effectuer : 2,5 m pour une presse à manioc, 1,1 m pour un tamis … Toutes les tiges seront ainsi coupées à la même longueur à l'aide d'un couteau, en incisant circulairement la tige afin de couper net la canne.

c. Gratter

Les tiges sont ou non grattées en fonction de l'ouvrage à réaliser. Si c'est un panier portant des motifs colorés, la couche chlorophyllienne recouvrant la tige devra être grattée à l'aide du dos de la lame d'un couteau. L'artisan prendra bien soin de laisser quelques centimètres non grattés à chaque extrémité de la tige. Pour certaines vanneries, comme la presse ou le tamis, les brins n'ont pas besoin d'être grattés.

d. Enduire de teinture les tiges grattées

Pour les teintures naturelles, sont utilisés : pour le rouge, du *roucou* (*Bixa orellana*, Bixacées) ; pour le noir, de la suie mélangée avec la pulpe gorgée de sève du liber de *bougouni* (*Inga* spp.[5], Mimosacées). On applique la teinture à la main sur la tige grattée.

e. Inciser

La partie laissée verte est incisée circulairement et légèrement, avec un couteau bien aiguisé, à 5-10 mm de l'extrémité la plus fine de la canne d'arouman. On crée ainsi une languette qui servira à tirer le brin utile.

f. Fendre

Les tiges sont fendues en deux puis en quatre. Pour initier le fendage, l'artisan utilise soit une croix grossière[6] faite de chutes d'arouman ou un couteau. Mais il est impératif de toujours fendre manuellement la tige d'un geste sec.

g. Enlever la moelle

La moelle interne de chaque partie de la tige est enlevée grossièrement à l'aide d'un couteau ou bien uniquement à l'aide des doigts. À cette occasion, on sépare encore en deux, avec les doigts, chaque brin.

h. Séparer le brin de la moelle

À l'aide de l'index, on sépare le brin utile du reste de la moelle en s'aidant de la languette incisée précédemment. Avec la main gauche, on tient le brin « déchet » et avec l'index de la main droite on tire vers soi le brin utile en exerçant une pression équivalente à celle exercée par la main droite afin d'obtenir un brin régulier et de ne pas le rompre. Cette action est la plus délicate de toutes et nécessite un long apprentissage, mais la réalisation d'un brin régulier est aussi conditionnée par la bonne exécution des étapes précédentes. On obtient ainsi un brin tressable et un autre brin qui est soit jeté soit utilisé par les jeunes pour l'apprentissage. Ce brin est cassant.

i. Ébarber

Avant de commencer le tressage de la vannerie, il faut ébarber certains brins à l'aide d'un couteau, afin d'éviter tout risque ultérieur de blessure par écharde. Une mâchoire de piranha aux dents effilées devait faire usage auparavant.

j. Tresser la vannerie

La récolte et la transformation de l'arouman ne nécessitent aucun outil spécifique, les deux seuls instruments utilisés étant des outils de base : la machette ou sabre d'abattis pour récolter la tige ; le couteau (ou le canif) pour couper, gratter et inciser la canne. Ceux-ci sont présents dans tous les foyers, la machette étant l'outil à tout faire des sociétés forestières tropicales. Notons qu'elle est le seul outil utilisé dans l'agriculture, le bâton à fouir n'étant utilisé, en Guyane, que pour planter le maïs chez les Wayana vivant sur le haut Maroni et les Wayãpi de l'Oyapock (P. Grenand, com. pers.).

[5.] Trois espèces principales du même genre sont utilisées pour le même usage : *Inga alba, Inga bourgouni et Inga pezizifera*.

[6.] Il s'agit d'un outil temporaire, fait de deux morceaux d'une tige d'arouman fendue, attachés entre eux.

Avant l'introduction du métal par les occidentaux, outre les haches en pierre, les outils destinés à couper étaient confectionnés avec des dents ou des végétaux comme le bambou. En Guyane, on utilise encore une demie mâchoire de pécaris pour raboter les arcs. Cet outil a sûrement dû être employé pour gratter la couche chlorophyllienne de la tige d'arouman. Des couteaux étaient aussi confectionnés en dents de poisson comme le piranha ou en dent de rongeur comme l'agouti[7] ou même en bambou refendu. Un voyageur ayant parcouru les Antilles au XVIIᵉ siècle rapporte l'usage, par les Caraïbes insulaires des petites Antilles, d'un couteau fait d'os et d'une dent d'agouti pour couper différents matériaux (Anonyme de Carpentras 1990). Ce même outil était utilisé par les Wayana du Brésil dans un passé récent (Van Velthem 1998).

Outre la coupe de la tige d'arouman, son incision et son éventuel grattage, toutes les autres opérations se réalisent à la main. Il en est de même pour la préparation de toutes les autres plantes employées dans la vannerie, faisant de cette activité un artisanat manuel par excellence.

La majorité des vanneries tressées avec les brins d'arouman le sont suivant la technique nattée, les brins plats et souples se prêtant bien à cette technique, contrairement aux brins ronds de saule (*Salix* spp.) par exemple, plus raides, utilisés dans nos campagnes européennes et qui se prêtent mieux à la technique clayonnée. On touche là un point important en vannerie ; la forme du brin induit la plupart du temps une technique de tressage préférentielle. A. Leroi-Gourhan (1973) avait déjà largement souligné que la matière première influe sur la technique.

Dans le même ordre d'idées, certains vanniers confirmés considèrent que chaque espèce d'arouman se prête plus spécifiquement à tresser certains objets. De même, on ne récoltera pas des aroumans du même âge suivant la vannerie destinée à être tressée : une presse à manioc, un tamis ou un panier, par exemple. Une connaissance fine de l'écologie de la plante, de son cycle de croissance ainsi que de ses qualités techniques paraît nécessaire pour devenir un artisan confirmé. Non seulement les artisans apprécient les différences de qualités techniques entre les différentes espèces d'arouman mais ils reconnaissent aussi que le contexte environnemental influe sur les qualités de la fibre au sein d'une même espèce. On note là l'importance d'une identification précise prenant en compte toute la finesse des taxons vernaculaires. Seule une approche ethnobotanique permet de mieux appréhender ces taxonomies populaires.

VANNERIE ET MANIOC : OUTILS, LIEN SOCIAL ET MARQUEUR ETHNIQUE

La presse à manioc ou *couleuvre*

Le manioc (*Manihot esculentus*) est fondamental dans l'alimentation des sociétés forestières de Guyane. Cet arbuste de la famille des Euphorbiacées comprend deux grandes formes, le manioc doux et le manioc amer[8]. La variété douce domine le sud de l'Amazonie alors que la partie nord de cette région ainsi que le plateau des Guyanes sont dominés par la culture du manioc amer. Ce dernier, représenté par plusieurs variétés, reste majoritairement cultivé et consommé en Guyane, spécialement chez les Amérindiens.

De part son apport important en protéines, sa grande productivité, sa facilité de culture, sa résistance aux agents pathogènes (McKey et Beckerman 1993) et ses grandes capacités de conservation (Mowat 1989, Dufour 1993), le manioc amer constitue une ressource alimentaire de choix. Il contient beaucoup plus de calories par hectare que le maïs par exemple (Mowat 1989). Ses qualités organoleptiques et le grand nombre de recettes réalisables sont aussi des facteurs non négligeables pour les Amérindiens, comme le montre l'étude de D. Dufour (1993) chez les Tukano. Et, comme l'explique F. Grenand (1993), les aspects économiques, culturels et sociaux sont tout aussi déterminants dans le choix de cette racine comme aliment de base.

Il est extraordinaire que ces sociétés aient basées leur régime alimentaire sur une racine contenant de l'acide prussique, un poison mortel ; les populations amazoniennes mangeant le manioc amer demeurant les seules au monde à avoir basé leur alimentation sur un aliment toxique (McKey et Beckerman 1993, Grenand F. 1996a).

[7.] *Dasyprocta agouti* est un rongeur de la taille d'un lapin. Sa chair est couramment consommée en Guyane.

[8.] La distinction entre manioc doux et manioc amer est basée sur la concentration respective des deux tubercules en cyanure hydrolysable. C'est Koch (cité dans Dufour 1993) qui a établi le premier un barème : au dessus de 100 ppm de cyanure hydrolysable, par racine pelée, le manioc est dit amer et en dessous de ce chiffre, le manioc est dit doux.

Ainsi, pour rendre comestible ce tubercule, les Amérindiens ont mis au point, tout au long de leur histoire, des outils élaborés afin de transformer cette racine en aliments divers et variés. Car c'est tout un ensemble de transformations qui contribue à sa détoxication, la technique qui enlève le plus sûrement les toxines étant l'essorage grâce à la presse à manioc (Grenand F. 1996b). Cependant, c'est tout un processus, allant de la déshydratation au chauffage en passant par la fermentation, qui est nécessaire à la destruction quasi-totale des molécules toxiques (Mc Key *et al.* 1993).

Ce tubercule est hautement valorisé et présent, tous les jours, dans la vie des Amérindiens que ce soit sous forme de galette (*cassave*), de farine torréfiée (*couac*), de condiment (*tukupi* ou *couabio*), de boisson (*cachiri*) … Bien sûr, il existe pour chaque préparation de nombreuses recettes et chaque ethnie possède ses propres variantes[9].

Au cours de notre enquête auprès de toutes ces communautés, nous avons dénombré 90 formes de vanneries liées à la préparation et la consommation du manioc sur les 210 répertoriées, soit 43 % des formes totales, ce qui en fait la première catégorie d'usage[10].

---

[9] F. Grenand (1996a et b) a écrit deux articles détaillant différentes préparations du manioc chez les Wayãpi. D. Schoepf (1979) a lui décrit la cuisine wayana tandis que l'ouvrage édité par P. Erikson (2006) offre un large panorama de l'utilisation du manioc et d'autres plantes dans la confection des bières amazoniennes.

[10] Les populations étudiées classent les différentes formes de vannerie en fonction de leur usage. Ainsi, il nous a semblé pertinent de les présenter en fonction de leur usage. Les autres catégories sont : les vanneries pour le rangement et le stockage, les vanneries cérémonielles, les vanneries commerciales et enfin diverses vanneries telles que celles destinées à la pêche, le repos, le jeu …

Figure 6. *Tepisi* ou *Couleuvre* à manioc wayãpi.

Les formes principales sont des outils fonda-mentaux entrant dans le processus de détoxication du tubercule. La presse à manioc, nommée *couleuvre* en Guyane, en est la pièce maîtresse (fig. 6). Cet outil complexe est une invention des Amérindiens de Guyane (Métraux 1928, Nordenskiöld 1929). Il s'agit d'un long manchon ouvert à l'une de ses extrémités dans laquelle on fait entrer la masse préalablement râpée de la racine. Une fois suspendue à une poutre, on enfile un bâton dans l'anneau inférieur. Avec un poids, en s'asseyant sur le bâton ou en y calant une grosse pierre, la *couleuvre* est étirée, permettant ainsi d'exprimer l'eau de cette masse très humide contenant le suc toxique.

Les populations du bassin amazonien utilisent bien d'autres vanneries pour extraire les toxines de cette racine mais elles sont plus rudimentaires et moins efficaces. Il s'agit de nattes à torsion, utilisés par les Kayapo ou les Witoto, ou bien de tamis dans lesquels est pressée la masse comme chez les Trumai. Certains encore, comme les Tapirapé, la pressent directement entre leurs mains (Dole 1960).

Outre la *couleuvre* à manioc, les Amérindiens de Guyane tressent un grand nombre de formes toutes liées au complexe de production des aliments à base de manioc amer. Celles-ci se rencontrent à toutes les étapes, allant de la récolte à la consommation. Elles sont utilisées pour le transport (hotte ouverte ou en cloche), la trans-formation (presse, tamis), la cuisson (éventail à feu) et le séchage des galettes (natte). Ils tressent aussi divers paniers pour stocker la farine torréfiée ou la pâte de *cachiri*. Toutes ces vanneries constituent la majeure partie du mobilier présent dans les maisons amérindiennes avec l'inévitable hamac et les poteries (qui se font de plus en plus rares).

Des outils féminins tressés par des hommes

Toutes ces vanneries sont tressées par les hommes mais utilisées uniquement par les femmes, la transformation du manioc étant une tâche féminine comme en grande partie sa culture. Plus largement, en Amazonie, la vannerie est une activité majoritairement masculine, à part chez quelques peuples comme les Bororo (Lévi-Strauss 1964) ou les Yanomami (Biocca 1968). Par contre, les femmes utilisent en majorité les vanneries, 60 % des vanneries leur étant destinées. Les hommes, eux, ne se servent que de 20 % des formes recensées alors que les 20 % restants sont à l'usage des deux sexes.

Ainsi, les femmes reçoivent des vanneries de leur mari ou d'autres consanguins pour transformer le manioc mais aussi pour stocker des aliments comme les piments boucanés ou des légu-mineuses. Il existe aussi un nombre important de beaux paniers, toujours décorés, destinés à ranger le coton ou la pelote filée et son fuseau.

Les hommes ne possèdent que de rares vanneries comme divers coffres et coffrets afin de conserver leurs plumasseries, leurs colliers ainsi que divers autres petits matériels. Les chamanes, nommés *piaye* en Guyane, rangent dans leur coffre ou *pagara* leur hochet, leurs cigares ou tout autre ustensile cérémoniel.

Par contre s'ils n'utilisent que peu de vanneries, les hommes ont le monopole en Guyane du tressage des vanneries à usage domestique. Ce n'est que tout récemment que l'on a vu apparaître, dans la communauté palikur, des femmes tressant des formes modernes de paniers afin de les vendre aux touristes (Davy 2002). Mais elles ne tressent toujours pas les formes utilitaires, traditionnelles. Dans les sociétés amazoniennes, toutes les activités se répartissent en fonction du sexe, des interdits stricts les encadrant. Traditionnellement, tout homme ou femme doit maîtriser les tâches incombant à son genre. Il n'existe pas, en effet, de répartition des activités en fonction de castes ou de classes comme il en existe dans les sociétés andines, en Afrique ou en Asie.

Un artisanat valorisé socialement

Comme les technologues culturels le soulignent l'action technique est productrice de vie sociale (Digard 1979). Ainsi, en Guyane, la pratique de la vannerie est constitutive de l'identité masculine amérindienne, même s'il est vrai, que cela est de moins en moins le cas aujourd'hui. Il n'empêche qu'elle est toujours une activité valorisée et reconnue. On respecte et reconnaît un habile vannier à l'instar d'un bon chasseur. Si tous les hommes se doivent de maîtriser la confection des vanneries élémentaires, il existe néanmoins cer-taines personnes plus habiles que d'autres, sachant tresser un plus grand nombre de formes ou de motifs. Traditionnellement, il n'y a pas à propre-ment parler de spécialistes car potentiellement tous les hommes se doivent de connaître le maximum de formes et de motifs. Néanmoins aujourd'hui, en raison de la perte de la transmission du savoir, il existe un nombre grandissant d'hommes, souvent les plus jeunes, ne sachant plus tresser. C'est le cas des Amérindiens du littoral, des Kali'na, Palikur ou Arawak chez qui il existe des spécialistes tressant des vanneries et les vendant à des foyers dans lesquels l'homme ne sait pas ou ne peut plus tresser en raison d'un travail salarié, par exemple.

Par contre à l'intérieur de la Guyane, chez les Wayana ou les Wayãpi, il est encore rare qu'un père de famille ne tresse pas. Les hommes sachant tresser de nombreuses formes et connaissant un grand nombre de motifs sont reconnus, connus de tous et considérés avec respect. Le fait de bien maîtriser cet artisanat indique un homme connaissant bien sa culture sous tous ses aspects, car comme nous l'évoquerons plus loin, les motifs et les formes sont aussi des marqueurs ethniques. De plus, les motifs en représentant des animaux mythiques sont de véritables supports de la mémoire collective de ces populations.

Traditionnellement, tous les jeunes garçons regardent dès leur plus jeune âge les hommes de leur parenté tresser un tamis ou un panier. Puis vers l'âge de 8-10 ans, ils commencent à imiter leur père ou un oncle en utilisant des brins qu'on aura préparés à leur intention ou bien des brins de moindre qualité. On commence par apprendre à tresser les formes les plus simples comme les paniers ajourés à trois nappes enchevêtrées jusqu'aux vanneries les plus élaborées que l'on devra savoir confectionner lorsque l'on atteint l'âge du mariage. Ensuite tout au long de leur vie, les hommes les plus doués et les plus motivés affineront leur technique et augmenteront leur corpus iconographique. Néanmoins, traditionnellement, un jeune homme prêt à se marier se doit de savoir tresser les vanneries essentielles liées au manioc pour qu'une famille accepte de lui donner sa fille. Sinon, c'est l'un des rares prétextes que peuvent prendre des parents pour se récuser (Grenand P. 1982). Chez les Kali'na vivant dans l'embouchure du Maroni, sur le littoral, lorsqu'un garçon demandait la main d'une fille, ses futurs beaux-parents lui donnaient un brin d'arouman qu'il devait enchevêtrer pour en faire une boule afin de montrer sa dextérité. S'il n'y arrivait pas, le mariage pouvait lui être refusé. Cette pratique semble bien être tombée en désuétude.

Ainsi, le futur père de famille devait maîtriser la confection des vanneries essentielles à la production alimentaire du nouveau foyer, permettant ainsi à sa future femme de préparer les galettes ou la farine torréfiée de manioc afin de nourrir sa famille. Toute femme doit aussi préparer le fameux *cachiri* ou bière de manioc afin de pouvoir inviter des convives à des réunions de boisson plus ou moins importantes suivant le contexte.

En effet, dans le monde amérindien, la bière de manioc ou *cachiri* est un véritable ciment social présent dans la moindre des festivités (Grenand F. 1996a, Erikson 2006). Une femme se doit de préparer le *cachiri* pour participer à la vie du village et ainsi être reconnue. Nous sommes en présence d'un classique système de don/contre-don entre les vanneries données, par l'homme, à son épouse et celle-ci offrant en retour, à son mari et à sa parenté, la nourriture constituée de manioc, de viande ou de poisson. La femme aussi offre au nom de son mari du *cachiri* lors de fêtes de boisson, créant ainsi du lien social. Si le don est un moyen de rendre dépendant celui qui reçoit et donc d'induire un rapport de supériorité entre le donneur et le receveur, il entraîne également et surtout une certaine solidarité (Godelier 1996). Dans notre cas, l'homme en tressant tous les outils liés au complexe du manioc rend ainsi dépendant son épouse et affirme sa « supériorité » sur celle-ci en tant que producteur d'outils. Mais, il instaure également une solidarité et une complémentarité dans le foyer qui cimente durablement le couple.

Des formes et des motifs comme marqueurs ethniques

Nous avons montré que les vanneries sont importantes à l'intérieur du groupe ethnique, du village et de la famille, mais elles sont aussi un marqueur identitaire fort permettant de se distinguer des groupes voisins. Les objets comme facteurs de culture et de différenciation identitaire, voilà qui, somme toute, est un classique en ethnologie. Mais il est toujours intéressant d'observer et d'identifier des styles entre diverses ethnies à une période donnée et dans une zone géographique bien précise surtout si l'on s'intéresse aux mécanismes interculturels.

Comme l'ont montré plusieurs ethnologues (Butt Colson 1973, Grenand P. 1982, Gallois 1986, Dreyfus 1992, Whitehead 1993), le plateau des Guyanes précolonial connaissait de complexes réseaux d'échanges matériels, culturels et politiques entre les différents groupes ethniques, ceux-ci formant de grands ensembles culturels cohérents. D'après Dreyfus (1992, p. 80), ces réseaux constituaient « un espace politique de communication sociale et idéologique, un espace de circulation de biens, de personnes, de valeurs ». L'arrivée des européens brisa ces réseaux et morcela ces « chaînes de sociétés » (Amselle et M'Bokolo 1985, p. 34) en raison des dramatiques perturbations tant démographiques que sociales induites par cette colonisation (Hurault 1972). Il en résulta des fusions et des reconstitutions ethniques (Grenand et Grenand 1987, Grenand P. 2006) et de nouvelles constructions identitaires (Collomb 2000).

Ces réseaux d'échanges ont favorisé les emprunts techniques et culturels entre ces groupes tantôt alliés, tantôt ennemis. Car il existe, au-delà des spécificités que nous mettons en lumière, une indéniable unité culturelle de la civilisation matérielle des Guyanes (Roth 1924). De nombreux objets et formes d'objets sont largement partagés par les différents groupes ethniques vivant dans cette région. Et la vannerie en est une illustration particulièrement frappante. Néanmoins, de nombreuses différences de styles existent entre chaque groupe ethnique. Ainsi, un œil averti peut distinguer qui a confectionné tel ou tel éventail à feu, tel ou tel tamis. Mais inévitablement, des emprunts et des diffusions entre ces groupes qui se côtoient depuis des siècles se sont produits et nous pouvons trouver, dans toutes les communautés, des techniques ou des motifs provenant d'un groupe voisin. Les artisans compétents le savent très bien. Par exemple, les vanniers wayana nous ont tous signalé que telle technique d'attache du tamis aux baguettes a été empruntée aux Wayãpi. Ils détiennent une grande connaissance des styles des ethnies voisines. Quand on leur montre des photos de vanneries de groupes avec qui ils ont ou ont eu des échanges, ils identifient très vite leur origine.

Les différences de styles entre les groupes peuvent s'observer globalement suivant trois critères : la forme, les motifs et les techniques.

Pour illustrer le premier critère, prenons l'exemple des éventails à feu, chaque forme étant typique d'une ethnie (fig. 7). Les Palikur tressent un éventail carré, les Wayana un éventail rectangulaire et les Kali'na, Arawak, Emerillon et Wayãpi des éventails trapézoïdaux mais ayant chacun une forme différente. En plus d'une variabilité de forme, les fibres ne sont pas tirées de la même plante. Si les Wayana et les Palikur tressent leurs éventails en arouman, les quatre autres ethnies les tressent en pinnules de palmier du genre *Astrocaryum*. Et à l'intérieur de ce même genre, chacun des quatre groupes utilisent une espèce différente, sauf pour les Arawak et les Kali'na qui utilisent tous les deux l'*awara*, *Astrocaryum vulgare*.

Prenons un autre exemple, celui de la presse à manioc. Nous pouvons distinguer la provenance de cet outil en fonction de trois critères. Pour la confection de cet objet, les six ethnies utilisent toutes l'arouman. Par contre, chacune tresse de manière différente la forme de la partie sommitale ou tête, la forme de la bouche et celle de la partie basale ou queue. Ainsi en observant chacune de

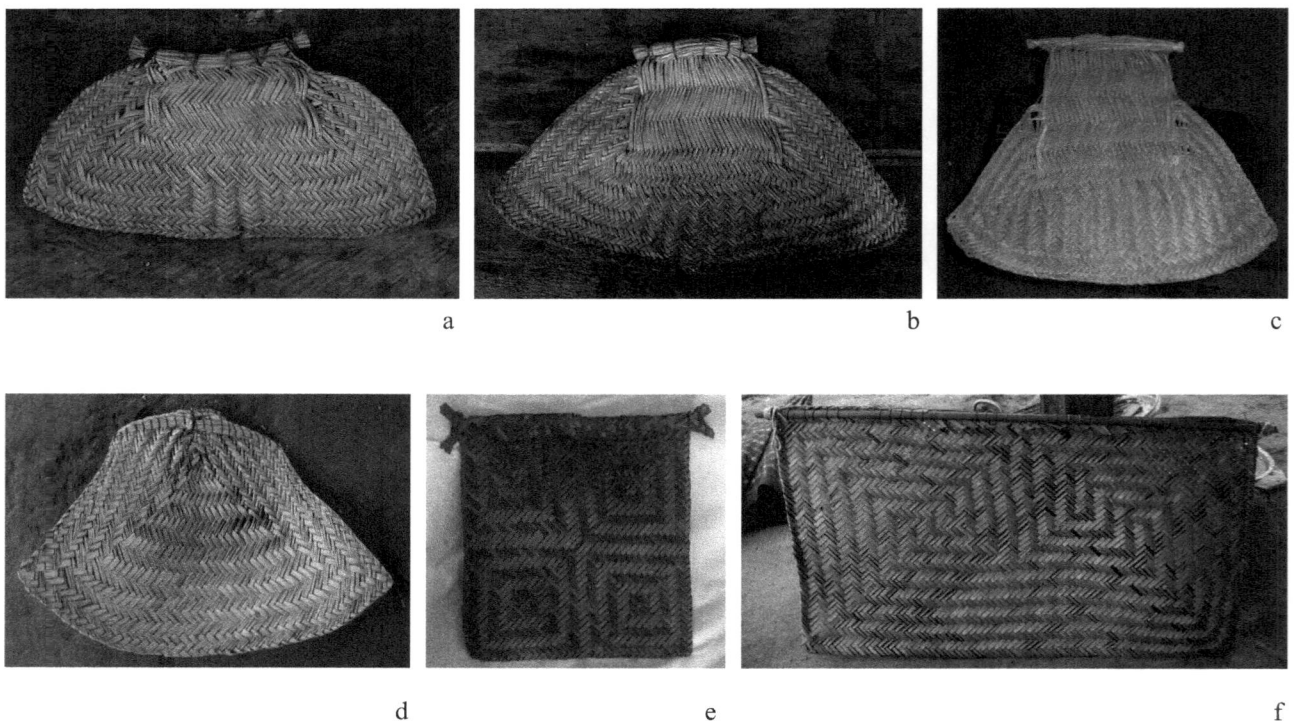

a   b   c

d   e   f

Figure 7. Les éventails à feu : a. *Wariwari* arawak ; b. *Woliwoli* kali'na ; c. *Tapekwa* teko ; d. *Tapekwa* wayãpi ; e. *Awagi* palikur ; f. *Anapamïi* wayana (photos de l'auteur).

ces parties, on peut identifier précisément l'origine de la vannerie. La couleuvre émerillon a la bouche droite, sa tête et sa queue sont très ouvragées et décorées d'un motif symbolisant le visage du jaguar. Tandis que la couleuvre wayãpi possède une bouche échancrée, la tête et la queue sont ornées de quatre petites nattes de brins noués avec du coton, nommées oreilles. Notons d'ailleurs que la grande diversité des techniques de tressage utilisées pour confectionner cette vannerie dans le plateau des Guyanes va dans le sens de l'hypothèse proposant cette région comme lieu de son invention.

On pourrait décliner ces distinctions qu'elles soient ténues ou évidentes pour toutes les autres formes de vannerie.

Un autre critère important marque l'appartenance ethnique. Il s'agit des motifs ornant paniers, hottes et tamis (fig. 8). Il existe, en effet, un nombre considérable de motifs encore connus. Nous en avons recensés environ 151 toutes ethnies confondues, chaque groupe ayant un corpus iconographique conforme à son *pattern* ethnique, même s'il est vrai qu'il existe certains motifs issus d'emprunts, partagés entre certaines ethnies. Au même titre que les formes et techniques, leur provenance est bien souvent connue des artisans. Néanmoins, les motifs commun à deux ou plusieurs ethnies ne sont pas très nombreux, la majorité du corpus iconographique étant typique de chacune d'elles puisqu'il est composé, en moyenne, de 60 % de motifs originaux. Ils sont constitutifs de leur identité et chaque culture amérindienne possède un mythe d'origine de ces motifs. Ces traditions orales content comment leurs ancêtres recueillirent et rapportèrent à leur communauté un répertoire graphique toujours utilisé jusqu'à présent pour décorer non seulement les vanneries, mais aussi les calebasses ou le corps.

a

b

c

d

Figure 8. Motifs de vannerie : a. *Matawat*, motif wayana représentant une chenille mythique bicéphale ;
b. *Wano pengapo,* motif kali'na représentant les alvéoles d'une ruche d'abeille ;
c. *Alamari* et *kupipi*, motif kali'na représentant l'anaconda mythique *alamari* et la grenouille *kupipi* ;
d. *Masuwili*, motif wayãpi représentant une hirondelle (*Hirundo rustica*, Hirundinidés) (dessin de Laurence Billault).

Les Wayana et les Wayãpi ont recopié leurs motifs sur le corps d'un anaconda mythique (Van Velthem 1998, Gallois D. 2002), les Palikur et les Teko sur celui d'une femme vautour à deux têtes (Davy 2006) et les Kali'na sur la carapace de la tortue de la femme lune (De Goeje 1955) ou, selon d'autres, sur un serpent géant. Ainsi, tous ces peuples ont recopié leur iconographie sur le corps d'animaux mythiques, souvent monstrueux selon nos critères, mais tout simplement hors du commun pour les Amérindiens. Remarquons d'ailleurs que dans l'univers amérindien les figures de l'anaconda ou du boa surnaturel « apparaissent en maîtres originels de tous les motifs décoratifs employés dans les peintures corporelles, dans les nattages des paniers et le tissage » (Lagrou 2005). Pour ces peuples, les motifs ont été donnés une fois pour toute et depuis ils les reproduisent à l'identique. Faisant partie intégrante de leur identité, ils leur permettent de se distinguer les uns des autres.

En tressant, on se remémore les mythes d'origine, comme chez les Yekuana du Venezuela où tresser est un moyen de méditer, surtout pour mieux comprendre les mythes (Guss 1989). De même, en Guyane, les motifs sont des supports permettant de se remémorer les histoires du passé. Ils sont des moyens mnémotechniques véhiculant le souvenir des récits d'antan et des entités extraordinaires qui les peuplaient. En effet, ils représentent de nombreuses entités surnaturelles ainsi que des animaux ou parties d'animaux vivant dans leur environnement. Ils permettent également de per-pétuer le lien entre les vivants et les ancêtres fondateurs d'une part, entre les vivants et leur environnement, d'autre part.

Les motifs phytomorphes, stellaires ou anthropomorphes sont eux relativement rares. Ainsi les paniers sont décorés de tout un bestiaire mythique comme des chenilles monstrueuses à deux têtes, des anacondas, des jaguars ou autres bêtes anthropophages. Les rares représentations de végétaux sont souvent liées au chamanisme comme, par exemple, la fleur de piment (plante utilisée dans les initiations des chamanes kali'na), la branche de l'arbre *wahusi*, *Virola surinamensis* (arbre dont l'écorce est hallucinogène[11]), etc. L'arouman est aussi présent chez les Wayana et les Kali'na avec le motif qui représente ses racines. Ces corpus iconographiques sont donc porteurs de

signifiant mais aussi apportent une remarquable touche esthétique aux objets les portant.

Mais tous ces motifs ne seraient-ils pas aussi utiles pour « habiller » les vanneries et les rendre, ainsi, civilisées comme l'a montré Guss (1989) dans le cas des Yekuana ? À l'instar des peintures faciales caduveo (Lévi-Strauss 1955) ou des ornements corporels matis (Erikson 1996), ces tatouages sur la « peau » des vanneries socialisent ces objets.

Ainsi quand un artisan wayana tresse une *couleuvre* à manioc (*tinkii*) ou un panier décoré d'un motif chenille à deux têtes (*matawat*), il confectionne non seulement un outil pour détoxiquer le manioc ou un panier pour que sa femme y range son coton, mais en plus il tresse des vanneries éminemment wayana. Il affirme ainsi son identité wayana en reproduisant un objet et un motif que son père et son grand-père tressaient avant lui[12].

Nous venons de voir que la vannerie en plus de son indéniable fonction utilitaire tient une place importante dans la vie sociale amérindienne en tant qu'outil social et marqueur ethnique. Une activité aussi importante dans ces sociétés est, comme on pourrait s'en douter, encadrée par des règles faisant partie d'une explication du monde dans laquelle s'inscrivent des mythes ainsi que des us et coutumes liés à la vannerie et aux plantes qui les constituent.

TRESSER LE MONDE : SYMBOLISME ET VANNERIE

Aux origines de la vannerie

De nombreux ethnologues, tenants de la technologie culturelle (Leroi-Gourhan 1973, 1992, Lemonnier 1986) ou de l'ethno-esthétique (Ribeiro 1989), ont montré que l'on ne peut séparer, dans les sociétés à tradition orale, la culture matérielle de la culture au sens large et des représentations qu'elle véhicule.

Comme nous venons d'en discuter, la vannerie tient une place importante dans les sociétés amérindiennes de Guyane. Et, à l'instar d'autres activités essentielles comme l'agriculture ou la

---

[11.] Le rapprochement est ici hypothétique car ce n'est que dans l'ouest des Guyanes et de l'Amazonie que cet usage est connu (P. Grenand, com. pers.).

[12.] Dans notre thèse nous montrons, de façon plus détaillée, la stabilité du répertoire graphique des peuples amérindiens de Guyane française.

poterie, quasiment tous les groupes ethniques connaissent un mythe expliquant comment les hommes apprirent la vannerie ou comment ils connurent les motifs les décorant. Dans tous ces récits la métamorphose, thème récurent en Amazonie, est omniprésent. Par exemple, selon la tradition, les Palikur, peuple arawak de l'embouchure de l'Oyapock, ont connu la vannerie grâce à un ancêtre qui, s'étant marié avec une femme oiseau cacique cul jaune (*Cassicus cela*), a appris à tresser et a rapporté ce savoir technique à son peuple. C'est depuis ce temps que les Palikur tressent leurs ouvrages comme le font ces oiseaux pour leur nid (Davy 2002).

Pour les Emerillon, peuple tupi-guarani du moyen Oyapock, c'est un héros culturel qui, après avoir été lui aussi marié à une femme oiseau mais cette fois-ci un vautour à deux têtes, rapporta à son peuple les techniques de vannerie ainsi que les motifs qui les décorent (Davy 2006).

Ainsi, dans la tradition orale de ces deux peuples, la vannerie fût apprise grâce au voyage d'un de leurs ancêtres, véritable héros civilisateur et à son union avec une femme d'un autre peuple. Rappelons que l'intermariage entre animal et humain pendant les temps mythiques est un thème récurrent dans la mythologie amazonienne (Grenand F. 1982), une période de métamorphose où nombre d'animaux et de plantes se sont transformés en humains et vice-versa, la frontière entre humains et non humains étant alors très perméable et encore mal définie.

Pour les Wayana, la vannerie existe depuis toujours. Après avoir fabriqué l'humanité en argile et l'avoir détruite, leur démiurge créateur *Kuyuli* tressa l'humanité et le monde avec des fibres d'arouman. Et, comme l'explique J. Chapuis « ce sont aujourd'hui les hommes qui en tressant l'arouman reproduisent la technique du créateur pour fabriquer le monde » (Chapuis et Rivière 2003). Ainsi pour ce peuple, comme chez les Kayabi du fleuve Xingu au Brésil (Ribeiro 1979), l'humanité a été tressée en arouman par le créateur. D'ailleurs, les Wayana disent que c'est pour cette raison qu'ils sont mortels car comme la fibre d'arouman, « l'homme se dessèche et meurt, si *Kuyuli* nous avait fait de pierre, on serait immortel ».

Si le thème classique de l'humanité fabriquée en terre cuite reste courant, il est, par contre, plus rare de rencontrer des peuples contant que leurs ancêtres ont été tressés avec des fibres végétales[13].

Ainsi pour les Wayana comme pour les Kayabi, vanniers réputés, leur condition humaine est intimement liée à l'arouman.

Ces divers exemples nous montrent un artisanat inscrivant son origine dans les temps mythiques, artisanat enseigné à l'humanité par des héros culturels ou bien directement appris d'un démiurge créateur lui même vannier. Ainsi tresser n'est pas une activité anodine, et plus spécifiquement tresser l'arouman, végétal constitutif du peuple wayana[14].

L'arouman, une plante pas comme les autres

Le mythe fondateur wayana montre le lien fort existant entre l'humanité et l'arouman. Tout au long de nos enquêtes chez les différents groupes amérindiens de Guyane, nous nous sommes aperçus que de nombreux interdits étaient liés au travail de cette plante. Mais aussi que les vanneries fabriquées avec ces fibres faisaient l'objet d'un traitement particulier.

Nous pensons, en effet, que si la vannerie est entourée de tant d'interdits, la matière première en l'occurrence, l'arouman, en constitue un facteur important.

De même que pour la vannerie, il existe des mythes d'origine concernant l'arouman. S'il ne manque pas dans la littérature ethnologique amazonienne de mythes d'origine de plantes utiles comme le manioc, les roseaux à flèches ou les lianes ichtyotoxiques, plus rares, sont par contre les mythes d'origine des plantes à vannerie.

Par exemple, chez les Palikur, deux espèces d'arouman ont été volées par des ancêtres à des animaux qui en étaient les détenteurs. L'une des espèces appartenait à l'agouti et l'autre, au tapir. C'est depuis ce temps que les Palikur utilisent ces deux espèces d'arouman, qu'ils nomment arouman de l'agouti et arouman du tapir (Davy 2002). Rappelons que ces deux animaux sont très présents dans la mythologie amazonienne et on les retrouve d'ailleurs dans des mythes arawak et karib comme possesseurs de l'arbre à nourriture qu'ils refusaient égoïstement de partager avec les Hommes (Lévi-Strauss 1964). De la même façon, pour les Palikur,

---

[13.] Les Ka'apor, groupe tupi-guarani vivant dans l'état du Marañón au Brésil, ont été créés à partir du bois d'arc (*Tabebuia serratifolia*) .

[14.] Il a d'ailleurs existé un clan proche des Wayana nommé *wamayana*, le peuple (*yana*) de l'arouman (*wama*) .

ces deux mammifères étaient les possesseurs égoïstes de ce végétal, utile pour tresser des outils indispensables à la production des aliments à base de manioc. Ils étaient donc propriétaires d'un végétal utile à la production de nourriture.

Chez les Teko, une espèce d'arouman est elle directement issue de la décomposition du corps d'un ancêtre. Ce personnage est mort à cause de la perte de son membre viril démesurément allongé au cours de péripéties cocasses et quelques peu salaces, dont les Amérindiens sont friands. Mourant, l'homme demanda à être enterré et sur son corps décomposé poussa l'arouman qu'utilisent depuis les Teko (Davy 2006).

Les Kali'na nomment la voie lactée, *waluma emali* que De Goeje (1910) traduit comme « le chemin où les ancêtres coupaient l'arouman ». On a là une relation directe entre le monde céleste, domaine des êtres créateurs, et l'arouman.

De plus, tous les artisans ont attiré notre attention en nous expliquant que l'arouman, comme nombre d'autres plantes, possède un esprit gardien. Chez les Kali'na, l'esprit de l'arouman est une petite fille portant une hotte sur le dos, pour les Wayãpi c'est un oiseau, *ulu*[15] ou perdrix toclo en créole (*Odontophorus guyanensis*). Pour les Teko, l'esprit de l'arouman est très fort et vit dans la boue dans laquelle cette plante pousse. Cet esprit ne serait-il pas la métamorphose de l'ancêtre sur lequel poussa l'arouman ?

Chez les Wayana également, un esprit de l'arouman existe et c'est grâce à lui, comme chez les Kali'na d'ailleurs, que l'on peut apprendre en rêvant[16] les motifs dont on décore les vanneries.

Il existe également des esprits dans certaines vanneries en arouman, celles-ci étant essentiellement des outils utiles au processus de transformation du manioc, tels que la presse à manioc et le tamis. D'après les Teko, l'esprit de la presse à manioc est particulièrement puissant et ressemblerait à un jaguar. D'ailleurs le pied et la tête de cet outil sont décorés d'un écusson de vannerie avec un motif symbolisant le visage du jaguar, animal prédateur

par excellence en Amazonie et qui est également lié au chamanisme (Chaumeil 2000).

Vannerie et métamorphose animale

Ainsi, les objets de vannerie ne sont pas de simples objets. Ils possèdent en quelque sorte une certaine vie. Ils sont d'ailleurs chez les Wayana et les Wayãpi issus de la métamorphose d'animaux monstrueux ou de parties du corps de ces bêtes (Van Velthem 2003). En effet, comme l'a souligné E. Viveiros de Castro (2004), « l'idée de création *ex nihilo* est virtuellement absente des cosmologies indigènes ». Ici les vanneries sont le fruit de transformations et non d'une création par un démiurge créateur. Par exemple, un panier wayana ajouré représente le jabot du vautour pape (*Sarcoramphus papa*). La tradition orale raconte comment des animaux des temps anciens se sont métamorphosés. Par exemple, dans un mythe wayana, l'éventail à feu se transforme en nid de guêpe, la presse à manioc en serpent géant (Hurault 1968).

De même, les Wayãpi racontent que des vanneries lancées dans la mer par *Yaneya* (démiurge créateur) se transformèrent en animaux pour dévorer des jaguars. La couleuvre se métamorphosa en anaconda, les éventails à feu en piranhas (Grenand F. 1982). Tous ces animaux ne sont pas anodins, ce sont en majorité des prédateurs dangereux et craints. Ainsi, les vanneries sont dotées d'un pouvoir ou, pour le moins, de forces bien réelles. On retrouve d'ailleurs un mythe récurent dans le plateau des Guyanes, contant comment un homme utilisa une vannerie, souvent une presse à manioc, pour tuer sa femme qui l'importunait. Ne serait-ce pas une métaphore de la puissance de l'esprit de la presse à manioc ? Rappelons qu'on la nomme *couleuvre* en Guyane, *couleuvre* signifiant anaconda en créole guyanais.

Ainsi, quand un homme tresse une vannerie, il tresse un animal mythique ou une partie de son corps, rendant possible « l'irruption du temps primordial et de ses composants dans la vie humaine actuelle » pour reprendre la formule de L. Van Velthem (2003). Ce serait ainsi le geste du vannier qui conférerait aux objets un certain pouvoir[17]. De

---

[15.] *Ulu* est également le nom de l'arouman dans cette langue. Il existe ainsi une synonymie entre la plante et son esprit gardien.

[16.] Le rapport étroit entre le monde onirique et la création artistique est largement présent dans de nombreuses parties du monde.

[17.] D'ailleurs, chez les Tilio, groupe karib, voisin des Wayana, un conte explique comment un homme ayant tressé une vannerie ornée d'un motif figurant un jaguar provoqua la métamorphose de cet objet en ce félin, montrant ainsi ses pouvoirs chamaniques et sa dextérité.

ce fait, de nombreuses règles se doivent d'être respectées afin de ne pas provoquer les esprits puissants présents dans ces objets et dans l'arouman lui-même.

Rites et interdits

Pour toutes les ethnies amérindiennes de Guyane, on ne doit pas brûler les anciennes vanneries ou les déchets issus de la préparation des brins d'arouman sous peine de perdre toutes les connaissances acquises dans cet artisanat. Ils doivent pourrir tranquillement, retournant progressivement à la terre.

D'après les Kali'na, tresser le soir est strictement interdit sinon, en guise de sanction, l'esprit de l'arouman peut rendre aveugle l'artisan et lui faire oublier les motifs.

De même des interdits stricts sont à respecter pendant la couvade, c'est-à-dire lorsqu'un couple vient d'avoir un enfant. Ce rite, très répandu en Amazonie, consiste pour les parents à respecter un nombre important de règles afin que le bébé, particulièrement vulnérable dans sa première année de vie, ne soit pas la proie des esprits de la forêt (Grenand F. 1984).

Ainsi, entre autres interdits, l'homme se doit de ne pas tresser l'arouman jusqu'à ce que le nouveau-né se déplace par lui-même. Si l'on tresse une presse à manioc ou tout autre ouvrage serré comme les ligatures de flèches, par exemple, l'enfant, particulièrement sensible aux mauvais tours des entités surnaturelles, risque de graves maladies comme des rétentions urinaires pouvant entraîner la mort selon les Wayãpi (Grenand F. *op. cit.*).

Ainsi, tout ce faisceau de relations entre l'arouman et les Hommes via la vannerie montre l'importance de cette plante pour ces peuples. Humanité tressée en arouman pour l'un, arouman issu de la décomposition d'un homme pour un autre, voie lactée vue comme un champ d'arouman pour un troisième … Et, pour tous, une place importante de l'esprit de l'arouman et de cette plante dans leur symbolisme. S'agissant d'un végétal possédant des esprits puissants qui confèrent aux objets tressés avec ses fibres une même puissance, on doit donc se concilier les faveurs de ces entités en respectant certaines règles.

CONCLUSION

Issue d'une longue tradition, la vannerie tient une place très importante dans les sociétés amérindiennes de Guyane française. Cet artisanat reste à la fois nécessaire pour confectionner des outils entrant dans le processus de transformation de la racine de manioc amer mais aussi pour tresser des paniers servant à conserver différents aliments, des hottes pour transporter du matériel et des produits de l'abattis, des nasses ou des paniers pour pêcher, des couronnes de plumes pour danser … En plus d'un aspect utilitaire, la vannerie est aussi un marqueur ethnique important, un vecteur identitaire véhiculé à la fois par des formes et un corpus iconographique fidèle au *pattern* ethnique. Quand un artisan tresse un panier orné de motifs, non seulement il confectionne un objet utile mais, en plus, il montre son appartenance à son peuple en reproduisant des canons esthétiques issus de sa tradition. Les motifs sont des marqueurs forts de l'identité ethnique et sont aussi des moyens de représenter non seulement l'environnement qui les entoure mais aussi de se remémorer les temps mythiques en tressant des animaux surnaturels les peuplant. De plus, soulignons, même si nous ne nous sommes pas attardé sur cet aspect, que l'aspect esthétique est aussi un critère important aux yeux des artisans (Davy 2007).

Afin de tresser tous ces objets, un grand nombre d'espèces végétales sont utiles, leur récolte dans le milieu naturel demandant une connaissance fine de l'écologie de celles-ci. Cela demande aussi une connaissance de leurs caractéristiques techniques. Si la vannerie nécessite un grand nombre d'espèces pour tresser les nombreuses formes, les outils utiles à la transformation de ces végétaux sont quasiment inexistants faisant de cet artisanat une véritable activité manuelle.

On peut ainsi considérer la vannerie des Amérindiens de Guyane comme un complexe culturel. En effet, en plus de son rôle utilitaire, elle engendre du lien social mais elle produit aussi des objets jouant un rôle important en tant que marqueur ethnique. De plus, les motifs qui l'ornent sont des supports de la mémoire collective permettant d'évoquer les histoires du passé et les entités qui les peuplent. Cet artisanat entièrement confectionné avec du végétal est aussi le fruit d'un usage et d'une connaissance fine de la nature et des plantes. La vannerie en tant qu'artisanat utile, social et symbolique est intimement liée à la matière première utilisée, tout particulièrement l'arouman, plante emblématique.

ADRESSE DE L'AUTEUR

Damien DAVY
IRD Orléans
5, rue du Carbone
45100 Orléans
France
davy@orleans.ird.fr

BIBLIOGRAPHIE

AMSELLE, J.-L. et M'BOKOLO, E., 1985. *Au cœur de l'ethnie : ethnies, tribalisme et État en Afrique*. Paris : La Découverte.

ANDERSON, R. L., 1979. *Art in Primitive Societies*. Englewood Cliffs: Prentice-Hall

ANONYME DE CARPENTRAS, 1990. *Un flibustier français dans la mer des Antilles 1618-1620*. *Première édition 1620*. Paris : Seghers.

BAHUCHET, S., 2000. La forêt matière. *In : Les peuples des forêts tropicales aujourd'hui, vol. II . Une approche thématique*. Bruxelles : Programme APFT II, 135-156.

BALEE, W., 1994. *Footprints of the forest, Ka'apor ethnobotany: the historical ecology of plant utilisation by an Amazonian people*. New York: Columbia University Press.

BIOCCA, E., 1968. *Yanoama*. Paris : Plon.

BLANC, L., FLORÈS, O., MOLINO, J.-F., GOURLET-FLEURY, S. et SABATIER, D., 2003. Diversité spécifique et regroupement d'espèces arborescentes en forêt guyanaise. *Revue forestière française*, n° spécial, 131-146.

BUTT COLSON, A., 1973. Inter-tribal trade in the Guiana Highlands. *Anthropologica*, 34, 1-67.

CHAPUIS, J. et RIVIÈRE, H., 2003. *Wayana eitoponpë : (une) histoire (orale) des indiens Wayana*. Guyane : Ibis Rouge.

CHAUMEIL, J.-P., 2000. *Voir, savoir, pouvoir, le chamanisme chez les Yagua de l'Amazonie péruvienne*. Genève : Georg.

COLOMB, G., 2000. Identité et territoire chez les Kali'na à propos d'un récit du retour des morts. *Journal de la Société des Américanistes*, 86, 149-168.

DAVY, D., 2002. *La vannerie et l'arouman, Ischnosiphon spp., chez les Palikur du village Kamuyene (Guyane française) : étude ethnobotanique d'une filière commerciale*. Master. Université d'Orléans.

DAVY, D., 2006. *L'artisanat de vannerie dans les communes du sud guyanais : état des lieux ethnoécologique et socio-économique*. Rapport final. Guyane : Mission pour la Création du Parc de la Guyane.

DAVY, D., 2007. La vannerie amérindienne en Guyane française, un patrimoine esthétique et culturel majeur. *In : L'art que cache la forêt*. Lambersart : catalogue de l'exposition, 58-63.

DE GOEJE, C. H., 1910. *Études linguistiques caraïbes*. Amsterdam: Johannes Müller.

DE GOEJE, C. H., 1946. *Études linguistiques caraïbes*. Amsterdam: North-Holland Publishing Company.

DE GOEJE, C. H., 1955. Philosophie, initiation et mythes des Indiens de la Guyane et des contrées voisines. *International archiv für ethnographie*, XLIV, 159.

DIGARD, J.-P., 1979. La technologie en anthropologie : fin de parcours ou nouveau souffle. *L'Homme*, 19 (1), 73-104.

DOLE, G., 1960. Techniques of preparing manioc flour as a key to culture history in Tropical America. *In:* A. F. C. Wallace, ed. *Selected Papers of the Fifth International Congress of Anthropological and Ethnological Sciences*. Philadelphia: University of Pennsylvania Press, 241-248.

DREYFUS, S., 1992. Les réseaux politiques indigènes en Guyane occidentale et leurs transformations aux XVIIᵉ et XVIIIᵉ siècles. *L'Homme*, 32 (22), 75-98.

DUFOUR, D., 1993. The bitter is sweet: a case study of bitter cassava (*Manihot esculenta*) use in Amazonia. *In:* C. M. Hladik, A. Hladik, O. F. Linares *et al.*, eds. *Tropical forests, people and food: biocultural interactions and applications to development*. Paris : UNESCO (13), 575-588.

ERIKSON, P., 1996. *La griffe des aïeux, marquage du corps et démarquages ethniques chez les Matis d'Amazonie*. Paris : Éditions Peeters.

ERIKSON, P., 2006. *La pirogue ivre : bières traditionnelles en Amazonie*. St Nicolas de Port : Musée français de la brasserie.

GALLOIS, D., 2002. *Kusiwa : pintura corporal e arte grafica wajãpi*. Rio de Janeiro : Museu do Indio (FUNAI).

GALLOIS, D. T., 1986. *Migração, guerra e comércio : os Waiãpi na Guiana*. São Paulo : FFLCH/USP.

GODELIER, M., 1996. *L'énigme du don*. Paris : Flammarion.

GRENAND, F., 1982. *Et l'Homme devint Jaguar : univers imaginaire et quotidien des indiens wayãpi de Guyane*. Paris : L'Harmattan.

GRENAND, F., 1984. La longue attente ou la naissance à la vie dans une société tupi (Wayãpi du haut Oyapock, Guyane française). *Bulletin de la Société Suisse des Américanistes*, 48, 13-27.

GRENAND, F., 1993. Bitter manioc in the Lowlands of tropical America: from myth to commercialization. *In*: C. M. Hladik, A. Hladik, O. F. Linares *et al.*, eds. *Tropical forests, people and food: biocultural interactions and applications to development.* Paris: UNESCO (13), 447-462.

GRENAND, F., 1996a. Cachiri : l'art de la bière de manioc chez les Wayãpi de Guyane. *In* : B.-B. M. C. et F. Cousin, eds. *Cuisines, reflets de sociétés.* Paris : Éditions Sépia du Musée de l'Homme, 326-345.

GRENAND, F., 1996b. Préparer et consommer le manioc chez les Wayãpi de Guyane. *In* : F. Cousin, ed. *Histoires de cuisines.* Paris : Éditions du Musée de l'Homme, 14-16.

GRENAND, F. et GRENAND P., 1987. La côte de l'Amapà, de la bouche de l'Amazone à la baie de l'Oyapock, à travers la tradition orale Palikur. *Boletim do Museu Paraense Emílio Goeldi*, 3 (1), 1-77.

GRENAND, P., 1982. *Ainsi parlait nos ancêtres. Essai d'ethnohistoire wayãpi.* Paris : ORSTOM.

GRENAND, P., 2006. Que sont devenus les Amérindiens de l'Approuague ? Réflexions autour d'une histoire peu documentée. *In* : S. Mam Lam Fouck et J. Zonzon Matoury, eds. *L'histoire de la Guyane, depuis les civilisations amérindiennes.* Guyane : Ibis Rouge, 105-126.

GUSS, D. A., 1989. *To weave and sing: art symbol and narrative in the south America rain forest.* Berkeley: University of California Press.

HURAULT, J., 1968. *Les Indiens wayana de la Guyane française, structure sociale et coutume familiale.* Paris : ORSTOM.

HURAULT, J., 1972. *Français et Indiens de Guyane.* Cayenne : Guyane presse diffusion éditeur.

KOELEWIJN, C. et RIVIÈRE, P., 1987. *Oral literature of the trio Indians of Surinam.* Foris publications.

LAGROU, E., 2005. L'art des Indiens du Brésil : altérité, « authenticité » et « pouvoir actif ». *In* : L. D. B. Grupioni, eds. *Brésil Indien.* Paris : Éditons de la Réunion des Musées Nationaux, 69-81.

LEMONNIER, P., 1986. The study of material culture today: toward an anthropology of technical systems. *Journal of Anthropological Archaeology*, 5, 147-186.

LEROI-GOURHAN, A., 1973. *Milieu et technique.* Paris : Albin Michel.

LEROI-GOURHAN, A., 1992. *L'homme et la matière.* Paris : Albin Michel.

LÉVI-STRAUSS, C., 1955. *Tristes Tropiques.* Paris : Plon.

LÉVI-STRAUSS, C., 1964. *Le cru et le cuit, Mythologiques I.* Paris : Plon.

MCKEY, D. and BECKERMAN, S., 1993. Chemical ecology, plant evolution and traditional manioc cultivation systems. *In*: C. M. Hladik, A. Hladik, O. F. Linares *et al.*, eds. *Tropical forests, people and food: biocultural interactions and applications to developement.* Paris: UNESCO (13), 83-112.

MÉTRAUX, A., 1928. *La civilisation matérielle des tribus Tupi-Guarani.* Doctorat. Université de Paris.

MOWAT, L., 1989. *Cassava and Chicha, bread and beer of the Amazonian Indians.* Aylesbury: Shire Publication.

NORDENSKIÖLD, E., 1929. The American Indian as an inventor. *The Journal of the Royal Anthropological Institute of Great Britain and Ireland*, 59, 273-309.

RIBEIRO, B. G., 1979. *Diário do Xingu.* São Paulo : Paz e Terra.

RIBEIRO, B. G., 1985. *A arte do trançado dos indios do Brasil : um estudo taxonômico.* Belem : Falangora.

RIBEIRO, B. G., 1989. *Indigenous art, Visual language.* São Paulo : Editora da Universidade de São Paulo.

ROTH, W. E., 1924. An introductory study of the arts, crafts and customs of the Guiana Indians. *In*: *38th annual report of the bureau of American ethnology.* Washington: Smithsonian Institution, 1-745.

SCHOEPF, D., 1979. *La marmite wayana : cuisine et société d'une tribu d'Amazonie.* Genève : Musée ethnographique.

VAN VELTHEM, L. H., 1998. *A pele de Tuluperê.* Belém : Museu Goeldi.

VAN VELTHEM, L. H., 2003. *O belo é a fera, a estética da produção e da predação entre os Wayana.* Lisboa : Assirio et Alvim.

VIVEIROS DE CASTRO, E., 2004. Exchanging Perspectives: The Transformation of Objects into Subjects in Amerindian Ontologies. *Common Knowledge*, 10, 463-484.

WHITEHEAD, N.-L., 1993. Ethnic transformation and historical discontinuity in native Amazonia and Guyana, 1500-1900. *L'Homme*, 126-128, 285-307.

www.ingramcontent.com/pod-product-compliance
Lightning Source LLC
Chambersburg PA
CBHW061003030426
42334CB00033B/3352